AF786101

L. Packer L. Prilipko Y. Christen (Eds.)

Free Radicals in the Brain

Aging, Neurological and Mental Disorders

With 68 Figures, and 22 Tables

Springer-Verlag
Berlin Heidelberg New York
London Paris Tokyo
Hong Kong Barcelona
Budapest

Packer, Lester, Ph.D.
Department of Molecular and Cell Biology
251 Life Sciences Addition
University of California at Berkeley
Berkeley, CA 94720
USA

Prilipko, Leonid, Ph.D.
Division of Mental Health
World Health Organization
ave Appia 27
1211 Geneva
Switzerland

Christen, Yves, Ph.D.
Fondation IPSEN pour la Recherche
Thérapeutique
30, rue Cambronne
75737 Paris CEDEX
France

ISBN-13: 978-3-642-77611-3 e-ISBN-13: 978-3-642-77609-0
DOI: 10.1007/ 978-3-642-77609-0

This work is subject to copyright. All rights are reserved, whether the whole or part of the material is concerned, specifically the rights of translation, reprinting, reuse of illustration, recitation, broadcasting, reproduction on microfilms or in other ways, and storage in data banks. Duplication of this publication or parts thereof is only permitted under the provisions of the German Copyright Law of September 9, 1965, in its current version, and a copyright fee must always be paid. Violations fall under the presecution act of the German Copyright Law.

© Springer-Verlag Berlin Heidelberg 1992
Softcover reprint of the hardcover 1st edition 1992

The use of general descriptive names, trade marks, etc. in this publication even if the former are not especially identified, is not to be taken as a sign that such names, as understood by the Trade marks and Merchandise Marks Act, may accordingly be used by anyone.

Product Liability: The publisher can give no guarantee for information about drug dosage and application thereof contained in this book. In every individual case the respective user must check its a accuracy by consulting other pharmaceutical literature.

27/3145/5 4 3 2 1 0 – Printed on acid-free paper

Preface

Free radicals appear to play a mojor role in many neurological (and non-neurological) dieseases. Both acute and degenerative disorders are thought to involve free radicals reactions in tissue injury (for a list see this book page 18).

This issue is very important for basic science and for therapeuthic approach as well. In order to review and discuss the place of free radicals in psychiatry and neurology, a joint meeting was initiated by the World Health Organization and the Fondation Ipsen in Paris on June 17–18, 1991. This book contains the papers presented during this meeting.

Although the papers included in this volume have been produced in the framework of WHO/Fondation Ipsen review of knowledge about free radicals and brain disorders, they express the views of the individual authors rather than a consensus of the participants at the meeting. They do not necessarily represent the decision or stated policy of the World Health Organiszation or the Fondation Ipsen.

The editors wish to express their appreciation to Mrs Mervaillie for the organization of the meeting and Mrs M.-L. Gage for her editorial assistance.

L. Packer, L. Prilipko, Y. Christen

Contents

Contributors

Bakalova, R. A.
Institute of Physiology, Bulgarian Academy of Sciences, Sofia 1113, Bulgaria

Binkove, B.
Institute of Experimental Medicine, Czechoslovak Academy of Science, 120000 Prague 2, Czechoslovakia

Bourre, J. M.
INSERM U. 26, Hôpital, Fernand Widal,
200 rue du Faubourg Saint Denis, 75475 Paris CEDEX 10, France

Ceballos-Picot, I.
CNRS URA 1335 – Laboratoire de Biochimie Génétique, Hôpital Necker-Enfants Malades, 149 rue de Sèvres, 75743 Paris, France

Doly, M.
Laboratoire de Biophysique, INSERM U 71,
Facultés de Médecine et de Pharmacie, Clermont-Ferrand, France

Dormandy, T. L.
Department of Chemical Pathology, Whittington Hospital,
London NW5 1RD, UK

Droy-Lefaix, M.-T.
IHB/IPSEN, Res. Lab., Le Plessis-Robinson, France

Goss-Sampson, M. A.
Membrane Biochemistry and Micro-nutrients Unit,
Institute of Child Health, 30 Guilford Street, London WC1N 1EH, UK

Halliwell, B.
Division of Pulmonary-Critical Care Medicine, UC Davis Medical Center,
4301 X St, Sacramento, CA 95817, USA

Hiramatsu, M.
Institute for Neurobiology, Okayama University Medical School, Okayama 700, Japan

Hwang, P. A.
Department of Pediatrics, Division of Neurology, Lagos University Teaching Hospital, P.M.B., Lagos, Nigeria

Javoy-Agid F.
INSERM U 289, Hôpital de la Salpêtrière, 47 Bd de l'Hôpital, 75013 Paris, France

Kagan, V. E.
Department of Molecular and Cell Biology, 251 Life Sciences Addition, University of California at Berkeley, Berkeley, CA 94720, USA and Institute of Physiology, Bulgarian Academy of Sciences, Sofia 1113, Bulgaria

Koynova, G. M.
Institute of Physiology, Bulgarian Academy of Sciences, Sofia 1113, Bulgaria

MacEvilly, C. J.
Membrane Biochemistry and Micro-nutrients Unit,
Institute of Child Health, 30 Guilford Street, London WC1N 1EH, UK

Mori, A.
Institute for Neurobiology, Okayama University Medical School, Okayama 700, Japan

Muller, D. P. R.
Membrane Biochemistry and Micro-nutrients Unit,
Institute of Child Health, 30 Guilford Street, London VC1N 1EH, UK

Ogunmekan, A. O.
Department of Paediatrics, Division of Neurology,
Lagos University Teaching Hospital, P.M.B. 12003, Lagos, Nigeria

Packer, L.
Department of Molecular and Cell Biology, 251 Life Sciences Addition, University of California at Berkeley, Berkeley, CA 94720, USA

Petkov, V. D.
Institute of Physiology, Bulgarian Academy of Sciences, Sofia 1113, Bulgaria

Petkov, V. V.
Institute of Physiology, Bulgarian Academy of Sciences, Sofia 1113, Bulgaria

Prilipko, L.
Division of Mental Health, World Health Organization, 1211 Geneva 27, Switzerland

Serbinova, E. A.
Department of Molecular and Cell Biology, 251 Life Sciences Addition, University of California at Berkeley, Berkeley, CA 94720, USA and Institute of Physiology, Bulgarian Academy of Sciences, Sofia 1113, Bulgaria

Sinet, P.-M.
CNRS URA 1335 – Laboratoire de Biochimie Génétique, Hôpital Necker-Enfants Malades, 149 rue de Sèvres, 75743 Paris, France

Sram, R. J.
Institute of Experimental Medicine, Czechoslovak Academy of Science, 120 00 Prague 2, Czechoslovakia

Staneva, D. S.
Institute of Physiology, Bulgarian Academy of Sciences, Sofia 1113, Bulgaria

Tyurin, V. A.
Institute of Evolutionary Biochemistry, Leningrad, USSR

Vaddadi, K. S.
University of Melbourne, Larundel Hospital, Plenty Road, Bundoora, 3083 Melbourne, Australia

Yokoi, I.
Institute for Neurobiology, Okayama University Medical School, Okayama 700, Japan

Free Radical Scavengers and Antioxidants in Prophylaxy and Treatment of Brain Diseases*

L. Packer

Increasing evidence indicates that probably most diseases at some point during their course undergo free radical reactions in tissue injury. In some cases, free radical reactions may be involved in multiple sites and at different stages of a chronic disease. Both acute and degenerative diseases are thought to involve free radical reactions in tissue injury.

An overview will be given of free radicals and the importance of antioxidant interventions with particular reference to the lipophilic antioxidant vitamin E and its interaction with water-soluble antioxidants in protection of biological systems.

Oxidants and antioxidants, the vitamin E cycle

A "free radical disease" is a disease in which free radicals are shown to be causative. Do aging, neurological and mental disorders involve any "free radical diseases?" There is now increasing evidence indicating that probably most diseases at some point during their course involve free radical reactions and tissue injury.

In some cases, free radical reactions may be involved in multiple sites and at different stages of a chronic disease. Stroke is an example; it may result from atherosclerosis of arterial blood vessels serving the brain. Free radical reactions may be involved in oxidation of low density lipoproteins (LDL), later in "foam cells" (macrophages loaded with damaged LDL), then in damaged endothelial cells and in lipid oxidation reactions involving smooth muscle cells together with other debris as the fatty streak forms, prior to an atherosclerotic plaque. A scheme illustrating some of the multiple sites involved is shown in Figure 1.

Hence, in the etiology and pathogenesis of neurological and mental disorders it is necessary to delineate the cause and effect relationship of free radicals to the temporal course of the disease. This information can be useful in planning therapeutić approaches and management strategies for patient care.

* Acknowledgments: Supported by NIH (CA 47597), ASTA MEDICA, and the Palm Oil Research Institute of Malaysia (PORIN)

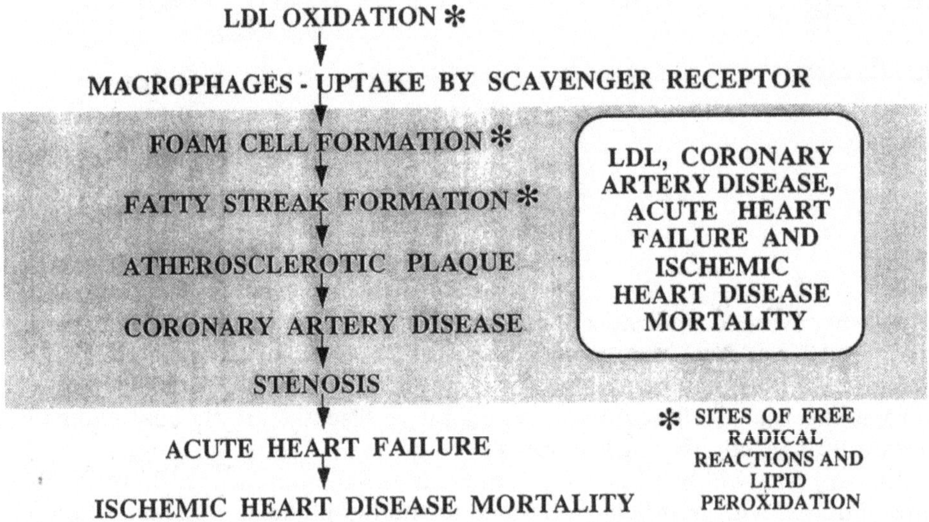

Fig. 1. Schemes showing some of the numerous sites at which free radical reactions initiated by lipid peroxidation and, therefore, protection by vitamin E, may be important in the formation of atherosclerosis and eventual ischemia-reperfusion injury, as in coronary artery disease and ischemic heart disease

The vitamin E cycle

Vitamin E is the name given to a group of naturally occurring lipid soluble antioxidants, the tocopherols and the tocotrienols, that are found in certain plant oils. Since the discovery of vitamin E in Berkeley in 1922 by H. M. Evans, when it was first described as an anti-sterility agent, many scientists and physicians have sought to elucidate its biochemistry, health benefits and clinical applications.

Vitamin E is the major, if not the only, chain-breaking anitoxidant in membranes, but its membrane concentration is very low, usually equal to or less than 0.5–0.1 nmoles/mg of protein (less than 1 molecule per 1000–2000 membrane phospholipids). The rate of lipid radical generation in membranes can be very high, about 1–5 nmoles/mg protein/min. Nevertheless, under normal conditions "rancidification", i.e. oxidation of membrane lipids and proteins, does not occur. Moreover, it is very difficult ro render animals deficient in vitamin E, and vitamin E deficiency is seldom found in adult humans. Hence, there must exist a remarkably efficient mechanism for permitting low concentrations of vitamin E to have such high efficiency in protecting membranes against damage and in supporting normal biological activity (Packer and Landvik 1989).

We hypothesize that these remarkable properties of vitamin E may be explained by its ability to be efficiently re-reduced from its free radical form (its form after quenching radicals) to its native state (Packer et al. 1989).

This occurs through the interactions between water- and lipid-soluble substances by non-enzymatic and enzymatic mechanisms, which regenerate vitamin E from its radical (tocotrienoxyl and tocopheroxyl radical back to tocotrienol and tocopherol, respectively; (Serbinova et al. 1991a). If true, this means that vitamin E interacts with water- and lipid-soluble substances either through chemical reactions or enzymatic pathways to drain away or accept their reducing power. Under conditions where these auxiliary systems act synergistically to keep the steady state concentration of vitamin E radicals low, the loss or consumption of vitamin E is prevented (Packer 1991; Hiramatsu and Packer 1991; Hiramatsu et al. 1990; Kagan et al. 1990a, b; Maguire et al. 1989; Mehlhorn et al. 1989).

This means that vitamin E will only be lost when these backup systems, either in the aqueous or membrane domains, become rate-limiting. At this point, vitamin E consumption will commence. The loss of vitamin E will be accompanied by increased rates of lipid and protein oxidation, destruction of membrane function, and inactivation of membrane enzymes and receptors. Thus, vitamin E not only has an antioxidant action, but acts as a biological response modifier, by modulating membrane-associated enzyme systems whose effects are amplified by the production of small molecular weight substances, like secondary messengers and products of the arachidonic acid cascade, which have profound effects at low concentrations on cell regulation and proliferation (Packer and Landvik 1989).

Oxidative damage in membranes will probably not occur while vitamin E is still present. Thus, accurate and sensitive measurements of vitamin E are of paramount importance, because it is only when vitamin E levels begin to fall and it disappears that the point will be reached when oxidative damage commences. The index of potential damage is high only when vitamin E is gone.

The concentrations of various other antioxidants which act in the cytosol to quench radical reactions or which serve to donate their reducing power to vitamin E may also fluctuate. The same may be true for membrane-associated redox couples such as ubiquinol/ubiquinones, or oxidized/reduced cytochrome c, which lowers the steady state concentrations of vitamin E radicals. Fluctuations in their concentrations are not as important as the concentration changes in vitamin E.

The thioctic acid/dihydrolipoic acid couple (TA/DHLA) is a unique antioxidant system (Packer 1991). Normally lipoamide exists as the cofactor of α-keto-dehydrogenases and is covalently bound in animals. Thus, its presence is at the level of micronutrient. However, TA, which can readily be fed to animals, has been found to confer protection in tissues and membranes against oxidative damage. This is believed to occur because TA, after it is absorbed, is reduced enzymatically or non-enzymatically and then becomes active as an antioxidant. It appears to interact directly with the membrane to reduce tocopheroxyl radicals (weak effect) or to reduce ascorbate, which in turn acts at membranes to reduce tocopheroxyl radicals (stronger effect). This unique mechanism of the TA/DHLA couple works in recycling of vitamin E both in membranes and in low density lipoproteins (LDL), where it acts to stabilize them. A redox cycle illustrating the inter-

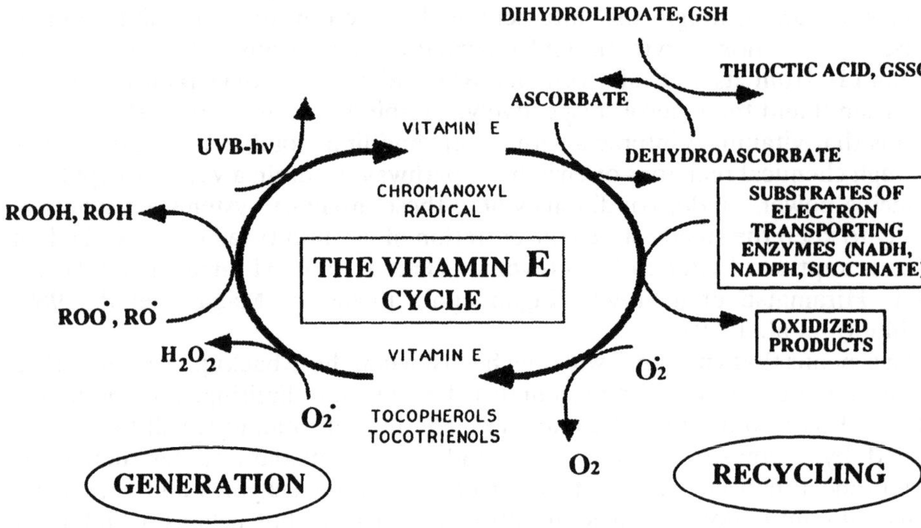

Fig. 2. The vitamin E cycle. Synergistic action of water- and lipid-soluble antioxidants. Non-enzymatic and enzymatic mechanisms for the regeneration of chromanoxyl radicals.
In membranes and human LDL, non-enzymatic mechanisms for the regeneration of the chromanoxyl radicals of vitamin E (toxotrienoxyl) and tocopheroxyl) include interaction with vitamin C (ascorbate). This pathway for regeneration of the radicals is bolstered by thiols that interact with ascorbate (dihydrolipoic acid, glutathione) and exert a synergistic effect in regeneration of the vitamin E radical. This pathway is present in both membranes and LDL. In electron transport membranes, enzymatic pathways for regeneration exist through substrate-dependent electron transport, where ubiquinol is a key component involved in vitamin E regeneration.
Both enzymatic and non-enzymatic pathways of vitamin E regeneration lower the steady-state (ESR signal intensity) of chromanoxyl radicals. This reduces the rate at which chromanoxyl radical-radical reactions destroy vitamin E. Hence, more efficient recycling is the key to the antioxidant efficiency of vitamin E in protection of membranes and LDL

actions between these antioxidants in membranes and LDL is shown in Figure 2.

Amplifying the vitamin E message – regulatory effects of vitamin E

Intracellular signaling and secondary messenger formation

It has been reported that protein kinase C (PKC), which is very important in intracellular signalling, may be down-regulated by vitamin E (Mahoney and Azzi 1988; Boscoboinik et al. 1991). This would have the effect of inhibiting cell proliferation. It has been suggested that these results may be relevant to the anti-cancer effects of vitamin E.

Recently, it has been recognized that mevalonate-derived intermediates of sterologenesis pathway suppress neoplastic growth (reviewed by Elson 1991). Among the substances that have been found to suppress mevalonate formation are products of tocotrienols (in red palm oil). Increased linoleic acid content in fat diets also suppresses tumor growth. The growth of several types of transplantable tumors is inhibited by tocotrienols, particularly the gamma isomer (Kato et al. 1985).

Thus, modulation of tumor growth rate by substances with polyunsaturated chains has potential anticancer activity. The extent that these effects are mediated through actions on prostanoid formation or on HMG CoA reductase activity is still unclear. These unique effects of the tocotrienols are another significant biological response modification effect of vitamin E.

Production of Cell Mediators by Arachidonic Acid Cascade

Phospholipase A2. Douglas et al. (1986) and van Kuijk et al. (1981) have shown that phospholipase A2 preferentially hydrolyzes peroxidized fatty acid esters of phospholipid membranes. Since lipid peroxides activate phospholipase A2, a decrease in lipid peroxide formation due to the presence of vitamin E will decrease phospholipase A2 activity. In this way, vitamin E modulates arachidonic acid release from membrane phospholipids and, therefore, arachidonic acid metabolism.

Lipoxygenase. Reddanna et al. (1985) and Bakalova et al. (1988) have demonstrated that vitamin E inhibits plant and mammalian 5- and 15-lipoxygenases. Fatty acid hydroperoxides necessary for activation of lipoxygenase can overcome the inhibition.

Cyclooxygenase. Halevy and Sklan (1981) reported that prostaglandin and hydroxyeicosatetraenoic acid (HETE) production from arachidonic acid in bovine seminal vesicles and kidney was inhibited by α-toxopherol. HETE production was inhibited less than that of prostaglandin. Thus it appears that vitamin E influences both the cyclooxygenase and lipoxygenase pathways; this modulation of arachidonic acid oxidation may have important *in vivo* implications.

Hypocholesteremic effects

There have been several reports that tocotrienols, the form of vitamin E with an unsaturated isoprenoid side chain, are hypocholesterolemic agents. Dietary studies feeding palm oil highly concentrated in α- and gamma-tocotrienol have demonstrated hypocholesterolemia; these effects are not exhibited by the corresponding tocopherols. Gamma-tocotrienol is particularly effective in this regard. This effect of tocotrienol to inhibit HMG CoA reductase activity, the first enzyme involved in the biosynthesis of cholesterol from mevalonic acid, was demonstrated by Qurseshi et al. (1986).

Coronary artery disease – Role of Lipophilic antioxidants

There is increasing evidence from epidemiological studies that the status of the human antioxidant defense system is important in assessing risk of chronic and degenerative diseases. Several such investigations in recent years have provided strong circumstantial evidence for beneficial effects of vitamin E. Low plasma levels of vitamin E are correlated with ischemic heart disease mortality and angina pectoris.

Ischemic Heart Disease (IHD) Mortality

The World Health Organization/Monica core study (c.f. Gey et al., 1991) was carried out with 20 populations in Europe. This is the most recent, extensive, and best controlled epidemiological investigation thus far conducted on risk factors involved in death from ischemic heart disease (IHD). There was a seven-fold difference in IHD mortality found. IHD correlated only 20 % with a combination of classical risk factors. Other indicators useful for prediction of risk were sought, and analysis of plasma antioxidants vitamins A and E was performed. As expected, a striking correlation was found betwen elevated plasma cholesterol and IHD mortality. This was found in 12 populations with varying cholesterol concentrations. The data showed a linear relationship between increasing cholesterol and IHD mortality. In eight populations, however, the cholesterol concentration did not differ. The strongest correlation was found between a low plasma vitamin E and A concentration and increased IHD mortality.

The gradient for increased IHD mortality ranged from the extreme northern countries, such as the Scandinavian countries, to the middle European and countries in southern Europe bordering the Mediterranean region. Seven-fold higher IHD was found in populations in the north as compared to the south. This was attributed to a higher proportion of yellow and green vegetables in the diet of people of the southern regions, a nutritional status that would be expected to provide much higher plasma concentrations of antioxidant vitamins.

Another striking correlation was made when the data from the 12 populations with the differing cholesterol concentrations were adjusted for cholesterol and lipid standardized factors. Here again, a highly significant correlation was obtained between lower risk to IHD mortality with higher plasma vitamin E and A levels.

Angina pectoris

Assessing the risk of angina pectoris in relation to plasma concentrations of antioxidants was the subject of a recent epidemiological study by Riemersma et al. (1991). In this study the hypothesis that plasma concentrations of antioxidant vitamins are related to the risk of angina was tested. The extent to which the risk of angina is independent of classic risk factors for coronary heart disease was evaluated. The factors which were evaluated were extensive, including age adjustment, habitual smoking, blood pressure, lipid composition, relative body weight, and also seasonal trends in plasma vitamin and antioxidant concentrations in samples from 6000 men aged 35 to 54.

Without adjustment for the various risk factors, low plasma concentrations of not only vitamin E, but also vitamin C and carotene, were related to an increased risk of angina pectoris in this male population. However, only for plasma vitamin E was the correlation significant after adjustment for the various risk parameters.

The correlation between angina and low plasma vitamin E concentration was the striking finding. Vitamin E was inversely related to the risk of angina, independent of the other antioxidants after adjustment for all of the classical risk factors. The adjusted odds ratio for angina between the lowest and highest quantities for vitamin E was 2.68 ($P = 0.02$). The investigators concluded that some populations having a high incidence of coronary heart disease may benefit from diets rich in natural antioxidants, particularly vitamin E.

Coronary artery disease – atherosclerosis

There now seems to be a clear picture emerging of how cholesterol and the LDL which carry them in blood are involved in the multistep process of coronary artery disease (CAD).

It was first shown in experimental animals (even before blood plasma lipoproteins were characterized) that atherosclerotic lesions occurred in animals fed diets high in cholesterol. Subsequently, animal studies showed that feeding with antioxidant vitamins such as vitamin E can prevent the occurrence of these lesions.

Polyunsaturated fatty acids are susceptible to lipid peroxidation and can form free radicals which can injure the endothelium, damage heart muscle cells, and provoke proliferation of smooth muscle. These processes have been shown to be inhibited by the antioxidant vitamins like C and E.

LDL can be oxidatively modified. Oxidation of the polyunsaturated fatty acids or cholesterol in LDL and oxidative modification of apoprotein B are involved in the subsequent recognition of damaged LDL by the scavenger receptor of macrophages. Extensive uptake of damaged LDL results in the formation of foam cells. Formation of these lipid-laden cells is among the first steps in the genesis of the atherosclerotic plaque. Foam cells later occur underneath the unbroken layer of endothelial cells, where they accumulate and form a fatty streak (which is comprised primarily of foam cells). Oxidation processes, particularly of lipids, occur in at least four different stages of atherosclerosis, starting from oxidative modification of LDL, oxidations in macrophages, formation of the fatty streak, and decomposition of smooth muscle, other cells and debris in the subendothelial space where the atherosclerotic plaque develops. Higher vitamin E concentrations would be expected to minimize or slow the rate of all of these processes. We speculate that this multifactorial location and site of action of vitamin E, not only as an antioxidant but also as a biological response modifier (see below), probably accounts for the correlation between low plasma levels of vitamin E and high risk of CAD.

Tocotrienols are more effcient antioxidants than tocopherols in membranes *in vitro* and in prevention of myocardial ischemia-reperfusion injury and oxidative damage to human LDL

Oxygen, lipid, and protein free radicals are implicated in cardiovascular injuries (myocardial ischemia-reperfusion, atherogenic peroxidation of LDL), and antioxidants have proven to be efficient cardiovascular protectors. d-α-Tocopherol and d-α-tocotrienol, two forms of vitamin E, are natural membrane antioxidants, having the same aromatic chromanol "head", but differing in their hydrocarbon "tail", d-α-tocopherol with a saturated and d-α-tocotrienol with an unsaturated isoprenoid chain.

Membranes – d-α-Tocotrienol had 40–60 times higher antioxidant potency in membranes than d-α-tocopherol due to the combined effects of three properties exhibited by d-α-tocotrienol as compared to d-α-tocopherol: 1) higher recycling efficiency from chromanoxyl radicals, 2) more uniform membrane bilayer distribution, and 3) stronger disordering of membrane lipids which renders interaction of chromanols with lipid radicals more efficient (Serbinova et al. 1991b). This further confirms that d-α-tocotrienol may have higher physiological activity than d-α-tocopherol under conditions of oxidative stress.

LDL – Both d-α-tocopherol and d-α-tocotrienol protected human LDL against oxidative modification induced by lipoxygenase, UV-irradiation, or azo-initiator of peroxyl radicals. ESR reveals chromanoxyl (d-α-tocopheroxyl and d-α-tocotrienoxyl) radicals in oxidatively stressed LDL. In the presence of ascorbate, the recycling efficiency for d-α-tocotrienol was higher than that for d-α-tocopherol.

Myocardial ischemia-reperfusion studies – We fed palm oil vitamin E (POE a gift of A. Gapor, PORIM) containing 45% d-α-tocopherol and 55% d-α-tocotrienol to rats to determine if it conferred protection against ischemia reperfusion injury (Packer et al. 1991). Animals kept on this diet for 45 days exhibited a pronounced cardiac mechanical recovery: after 40 min. of ischemia the hearts from POE animals recovered 90% of the control value, whereas hearts from the initial group recovered only 17% of the initial level.

POE provided for a full protection against ischemia-reperfusion-induced lipid and protein oxidation in the heart. d-α-Tocotrienol was predominantly consumed during the reperfusion period compared to d-α-tocopherol (41% of d-α-tocopherol was found in the heart after reperfusion and only 21% of d-α-tocotrienol was detected). d-α-Tocotrienol demonstrated higher *in vitro* NADH-, NADPH-, succinate-, and ascorbate-dependent recycling effciency from chromanoxyl radicals in heart membranes (mitochondria and microsomes).

This suggests that d-α-tocotrienol was the main antioxidant protecting myocardial membranes against oxidation in the course of ischemia-reperfusion injury.

New horizons in vitamin E research: requirements for vitamin E

The Vitamin E Paradox

How do we reconcile the great efficiency of minute quantities of vitamin E in membranes with the beneficial effects of vitamin E supplementation on human health seen in chronic diseases and acute clinical conditions?

A local, but vitally important, deficiency in the membrane content of vitamin E can arise rapidly. Exhaustion of vitamin E could occur under conditions where oxidative stress overloads the antioxidant defenses (recycling mechanisms) which prevent vitamin E from being lost. Under these conditions, vitamin E concentration will decrease to a point where molecular damage to lipids, proteins, and nucleic acids can be expected. Examples of such acute situations are ischemia-reperfusion injury, e.g., myocardial infarct or stroke and subsequent reoxygenation, hemorrhagic and other forms of shock, exposure to extremes of environmental pollution or irradiation, or anti-neoplastic drugs.

Replenishment of this localized vitamin E deficiency in membranes often occurs slowly, even in tissues in which vitamin E turnover is rapid. Since considerable time is required before the vitamin E content can be re-equilibrated, the conventional method of vitamin E loading by dietary supplementation may not be successful in preventing tissue injury.

It usually takes days to weeks to substantially increase the vitamin E content of membranes. Water-soluble substances which recycle vitamin E and which can be introduced in a matter of minutes may be helpful in increasing the effective concentration of the residual vitamin E in membranes or lipoproteins. The use of water-soluble recyclers or water-soluble forms of vitamin E which are more rapidly distributed, such as vitamin E phosphate or vitamin E succinate, may also be helpful. Thus, the concept of vitamin E recycling may prove to be the key to developing new strategies for the treatment of acute conditions involving oxidative damage, i.e., myocardial and cerebral ischemia-reperfusion injury.

Biokinetics and Tissue Absorption of Vitamin E

The kinetics of tissue absorption of the different forms of vitamin E is not well known. In humans, information is really very scarce. Even in animals the best available information are from feeding vitamin E (usually tocopherols) or injecting tocopherols and following kinetics of tissue absorption. From such studies, fast and slow turnover tissues have been identified. It has also been learned that certain tissues accumulate large amounts of vitamin E; these include adipose tissue, the liver and certain tissues with very high oxidative processes, such as adrenal.

Another important factor in tissue distribution of vitamin E isomers is discrimination of the isomer when it is transferred to lipoproteins. This occurs in the liver where vitamin E is absorbed (unless it first goes directly from chylomicrons to adipose tissue). In rat liver, an α-tocopherol binding protein has been identified; however, its specificity for the different tocopherols, alpha, beta, gamma, and delta, or for corresponding tocotrienols, is unknown.

In our laboratory, POE-fed rat tissues accumulate tocopherols and tocotrienols in many tissues, different skeletal muscle fibers, liver, and so forth if high concentrations are supplemented in the diet (at equivalent concentrations of tocopherol to tocotrienol). Hence, the fragmentary animal data that exist for rats or hamsters cannot be applied at this time with any fidelity to humans. Future biokinetic and tissue absorption studies may help to direct our attention to the biological functions of tocopherol and tocotrienol isomers.

Vitamin E Requirement

Health status, lifestyle, and environment exert marked influence on the requirements for vitamin E. Determination of requirements for vitamin E are complicated by variations in susceptibility of dietary and tissue fatty acids to peroxidation and the difficulty in demonstrating vitamin E inadequacies in healthy adults. Vitamin E requirements may vary five-fold, depending on dietary intake and/or tissue composition. High polyunsaturated fat intakes increase vitamin E requirements due to the increased peroxidative potential of body tissues.

Free radical-mediated damage has been implicated in cellular and extracellular changes that occur over time in the aging process and in development of chronic diseases. Vitamin E and other antioxidants prevent or minimize oxidative damage in biological systems. At what level the antioxidant defense should be to protect the body from the high free radical concentrations that are unavoidable at the present time is one of the many new horizons for vitamin E research.

Cardiovascular diseases and stroke

Principles of ischemia-reperfusion injury and interventions with antioxidants and free radical scavengers

Models of ischemia-reperfusion injury have been investigated in many tissues. Heart, kidney, intestine, and the brain are vital tissues where ischemia-reperfusion injury can occur. Both acute and chronic circulatory situations can arise, causing ischemia.

Ischemia

During long periods of ischemia, a cascade of events occurs, resulting first from the conversion of the tissue from oxidative to anaerobic metabolism.

These events include lowering of pH, inhibition of membrane transport systems, efflux and influx of ions (particularly calcium uptake), and conversion of xanthine dehydrogenase to xanthine oxidase due to a calcium-stimulated protease (calpain). All of these are believed to set the stage for generation of large amounts of reactive oxygen species (ROS) when circulation is restored.

Ischemia-reperfusion

Thus, when reperfusion commences, substrates for xanthine oxidase have accumulated (from the decomposition of adenine nucleotides) and, when the tissues are exposed to oxygen, large amounts of ROS form, including hydroxyl radicals, which can be detected by spin-trapping techniques in the perfusate following reperfusion. In myocardial reperfusion models, most of the burst of oxygen radicals occurs within the first five minutes. It may take many more minutes before the production of ROS falls to the control levels. During this period, it can be expected that ROS would initiate lipid and protein oxidation. Indeed, evidence can be obtained from a variety of assays for lipid peroxidation (conjugated dienes, intrinsic fluorescence from accumulation of end products of lipid peroxidation, increased thiobarbituric acid reactants, etc.) that lipid peroxidation has occurred. Also, a variety of methods have demonstrated protein oxidation. Most notably, the appearance of protein carbonyl groups occurs when metal catalyzed reactions occurring near the active site of proteins and enzymes result in oxidation of amino acid residues to carbonyls. These increase markedly in the heart and in the brain during the period of ischemia-reperfusion. Not very much change occurs during perfusion or perfusion plus ischemic period.

In the case of the heart, loss of mechanical activity and its recovery, loss of high-energy phosphates and their recovery, and leakage of enzymes specific to the heart to the perfusate are found in the circulating perfusate.

It can be expected that crucial enzymes in ischemic-reperfused tissues will be damaged. In heart, one enzyme that has been identified to be oxidatively damaged, and which is important in energy metabolism, is creatine phosphokinase. This is inactivated and leaks from the tissue, both in its active and in its inactive form.

Are these principles of ischemia-reperfusion injury common for other tissues, such as the brain? Model systems of cerebral ischemia indicate that this is the case. The gerbil has been widely used as a model system because in this small animal, the two carotid arteries which serve the brain are readily accessible and can be occluded.

The mongolian gerbil has no *circulus arterious cerebri* (Willis). Willis's circle is anatomically composed of right- and left-arteria (A.) cerebri posterior, A. communicans posterior, A. carotis interna, A. cerebri anterior and one A. communicans anterior. Because of this arterial circle system located at the basal position of the brain, ligation of one side of A. carotis interna hardly induced ischemia in the brain of an experimental animal except for the

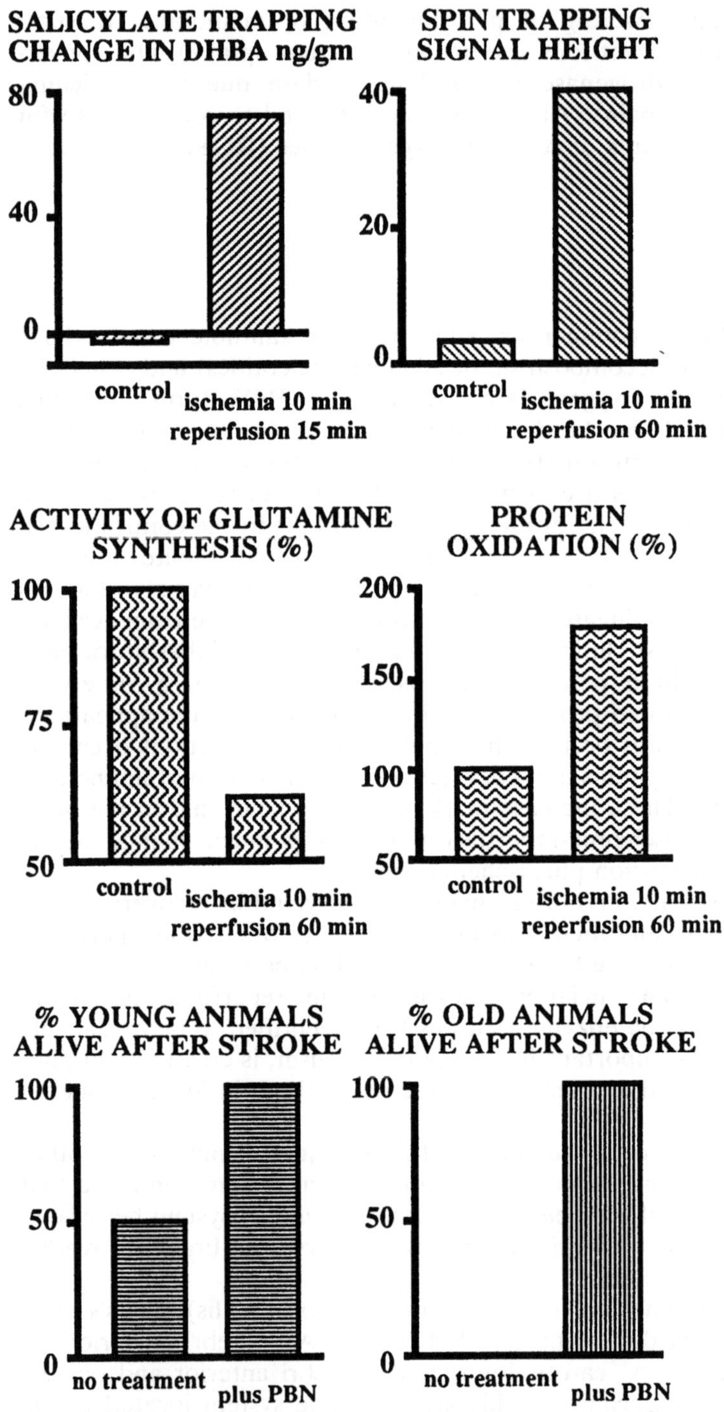

Fig. 3. Mongolian gerbil brain ischemia-reperfusion injury – Evidence for free radicals and oxidative damage (Floyd 1990)

mongolian gerbil. Hence the gerbil is an excellent model for preparing experimental ischemia.

Global ischemia to the brain can thus be introduced for varying periods of time and under controlled conditions. Such experiments (c.f., Floyd 1990 and Fig. 3) have revealed that increased concentrations of hydroxyl radicals, shown by the spin-trapping technique and also by salicylate oxidation, occur. In addition, protein oxidation, as determined by accumulation of carbonyl groups, occurs.

In the case of the brain, an enzyme that is specifically inactivated is glutamine synthetase. This causes glutamate to accumulate in brain, thus activating NMDA receptors, hence identifying the free radicals that occur during the ischemia-reperfusion phase in recovery from cerebral ischemia as being a crucial event in contributing to brain damage. This event clearly precedes activation of NMDA receptors; thus a complex management strategy including drugs or interventions that inhibit or quench free radicals will be required for treatment of such conditions, as is the case for myocardial infarction and heart failure.

These cerebral ischemia-reperfusion experiments also have demonstrated profound effects of free radical scavengers on recovery of the animals. For example, in old animals the death rate is high after cerebral ischemia reperfusion, whereas a higher proportion of the younger animals survives. Treatment with spin-trapping agents, which presumably trap radicals generated *in vivo*, has remarkable effects in preserving both young and old animals from death by stroke (Fig. 3).

Neurological behavior is also affected very strikingly by treatment with high concentrations of spin-trapping agents, which results in improved neurological function (c.f. Carney et al. (1991) and Fig. 4). Furthermore, spin-trap treatment decreases the extent of formation of protein carbonyls and inactivation of glutamine synthetase activity. After removal of the spin-trapping treatment, protein carbonyls and inactivation of glutamine synthetase activity reappear, strongly implying that these changes are caused by generation of radical species that react with the spin-trap. Behavioral changes are mainly seen in older animals (retired breeders) where the spin-trapped α-(4-pyridyl-1-oxide)-N-t-butyl-nitrate (PBN) is very effective in improving behavioral activity (Fig. 5). Hence, the brain is much more susceptible to damage due to exposure of ischemia in old compared to young animals.

Interventions with Antioxidants and Free Radical Scavengers in the Management of Ischemia-Reperfusion Disorders

Figure 6 shows general mechanisms of ischemia-reperfusion injury and antioxidant protection in different tissue model systems. During ischemia, the stage is set for the production of ROS during the early phases of reperfusion. Nevertheless, the endogenous antioxidant defeneses, both in the tissue cells and in the blood, are ineffective in protecting against ischemia-reperfusion-induced damage.

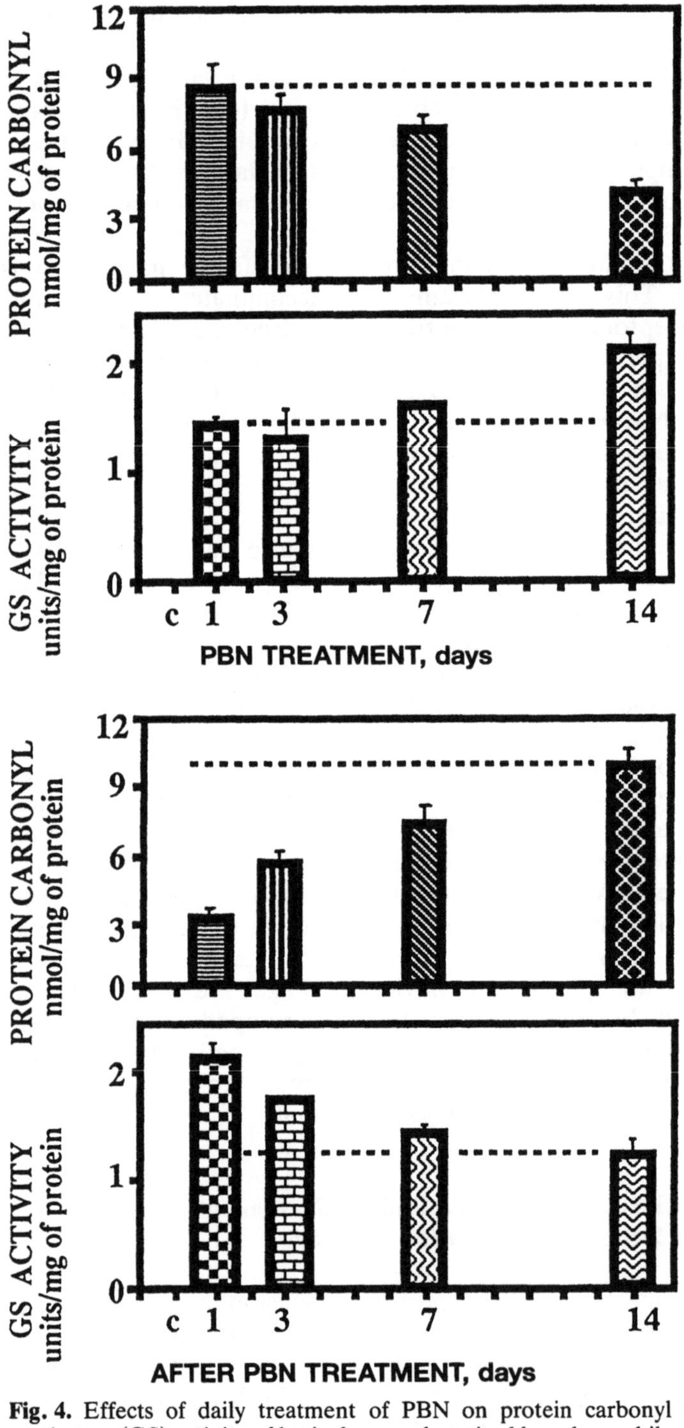

Fig. 4. Effects of daily treatment of PBN on protein carbonyl content and glutathione synthetase (GS) activity of brain from male retired breeder gerbils, and return to the original status after the last PBN dose (Carney et al. 1991)

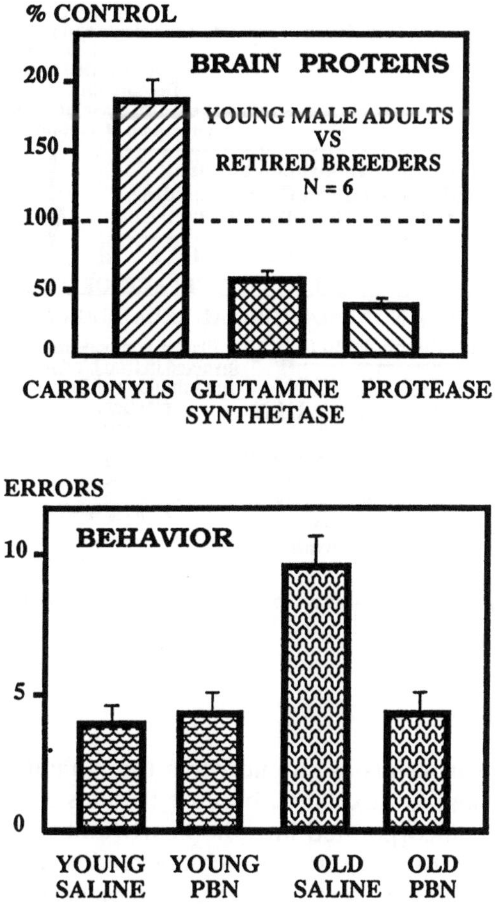

Fig. 5. Age-related changes of the protein carbonyl content, GS activity, and neural protease activity in brain of mongolian gerbils, and effects of aging and PBN treatment on radial arm maze patrolling behavior in gerbils (Carney et al. 1991)

However, a wide variety of radical scavengers and antioxidant interventions are effective if added externally or if the animals diet is previously supplemented with these agents. Exogenously added SOD and catalase frequently exert protection against ischemia-reperfusion injury. Likewise, desferrioxamine, the iron chelator, exerts protection if given into the circulation. Spin-trapping reagents in high concentrations can also be effective. Xanthine oxidase inhibitors, such as allopuranol, sometimes are effective, suggesting that xanthine oxidase contributes to the oxidative damage in these cases. Reduced sulfhydryl reagents, such as mecapto propionyl glycine, have been shown to be effective, suggesting that oxidation of critical sulfhydryl groups may be an important factor in the injury. Radical scavengers like L-propionyl carnitine appear to offer protection in ischemia-reperfusion, both by contributing short-chain fatty acids which can be used as fuel (this is especially relevant in myocardial ischemia-reperfusion) and by metal (Fe^{2+}) chelation. Various forms of vitamin E (the tocotrienols and tocopherols) are also effective in protecting against ischemia-reperfusion injury, but in this case,

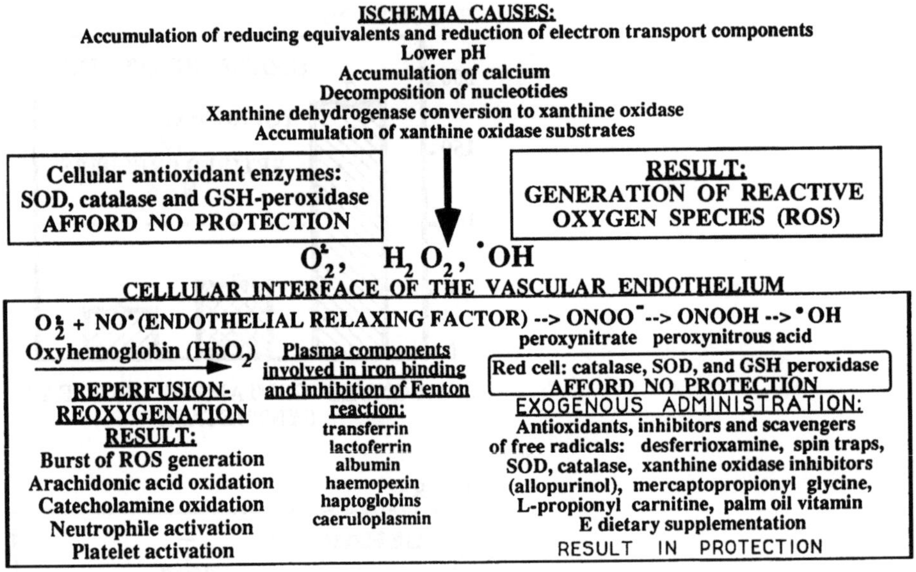

Fig. 6. Mechanisms of ischemia-reperfusion injury and antioxidant protection. (Adapted from Gutterdige and Halliwell 1990)

best results are found when the animals diet is previously supplemented with these antioxidants, because it takes a long time for lipophilic antioxidants to be incorporated into the heart (and also the brain).

Synthetic thiols, which are readily absorbed into the system, also appear to exert a protective effect. One example is thioctic acid (the oxidized form of lipoic acid) which, when ingested, becomes reduced by tissues and then offers protection in myocardial ischemia-reperfusion injury model systems.

Thus endogenous antioxidant defenses are ineffective whereas exogenous ones and radical scavengers are effective (c.f. Gutteridge and Halliwell (1990). The reason for this may be that penetration deep into the affected tissues cannot occur by the blood antioxidants and that greater protection is needed at the surface of the vascular endothelium, which may be the crucial area where damage occurs. This would explain why certain large molecules like SOD and catalase, which would not be expected to penetrate tissues, could offer protection in model systems.

Damage may also be exacerbated by the influx into damaged tissue of circulating leukocytes (polymorphonuclear leukocytes, PMN) when activated by ROS forms, exacerbating the degree to which tissue damage occurs in ischemia-reperfusion. The contribution in various model systems to ischemia-reperfusion, by activation of xanthine oxidase or activation of leukocytes or the combination of both, is an area of current interest and investigation. The extent to which these are important in ischemia requires further clarification and probably differs in different tissues and species.

Brain diseases and disorders involving free radicals-consensus of round table discussion

A round table discussion was held on these points as they relate to aging, chronic diseases such as stroke, mental disorders such as Alzheimer's disease and schizophrenia, and neurological disorders such as Parkinson's disease (Halliwell 1989) and epilepsy. We hypothesized that if the involvement of free radicals in the pathophysiology of these conditions could be more accurately elucidated, we could suggest and develop insights on how to design more accurate management strategies through antioxidant interventions.

At the beginning of the round-table discussion, Nordmann presented experimental data suggesting that acute as well as chronic ethanol administration elicits an oxidative stress in some parts of the central nervous system (Nordmann 1987; Rouach et al. 1987; Nordmann et al. 1991). This was shown in rat cerebellum by an increase in the decompartmentalized low-molecular-weight iron derivatives, a decrease in the main antioxidants (especially α-tocopherol, ascorbate, and selenium) and an increased susceptibility to lipid peroxidation. These disturbances may contribute to the ethanol-induced damage to the CNS, at least at the cerebellar level.

The high points of the round table can be divided into four parts.

The first part was devoted to the list of neurological diseases in which free radicals appear to play a role. All the diseases which had been considered on the preceding day during specific presentations were, of course, included (aging, ischemia, trauma, Parkinson's and Alzheimer's diseases, Down's syndrome, vitamin E deficiency, tardive dyskinesia occurring during treatment of schizophrenia patients with neuroleptics). In addition to these pathological conditions, already considered during the first day of the meeting, other neurological disorders were added to the list, appearing in Table 1.

Among very sensitive models of neurological damage by free radicals, oxygen-induced injuries of the retina were considered.

During the second part of the round table, the methods used for assessing the implications of free radicals during neurological diseases were extensively

Table 1. Brain diseases and disorders which involve free radicals (review in Jesberger and Richardson 1991)

1. Parkinson's disease	13. Progeria
2. Alzheimer's	14. Werner's syndrome
3. Stroke	15. Cocaine syndrome
4. Alcoholism	16. Vitamin E deficiency
5. Epileptic seizures	17. AIDS (HIV infection)
6. Head trauma (brain)	18. Oxygen inhalation therapy
7. Retinal damage	19. Shock
8. Spinal cord damage	20. Brain edema
9. Demyelination (multiple sclerosis)	21. Tardive dyskinesia
10. Schizophrenia	22. Inflammatory diseases
11. Dementia	23. Shock-brain damage by hyperoxia and hyperbaric oxygen
12. Down's syndrome	

discussed by the participants. The "classical" tests, i.e., determination of the thio-barbituric acid reacting substances and of conjugated dienes, were criticized – as they have been in many previous meetings! More specific techniques, using HPLC for the determination of lipid hydroperoxides and higher aldehydes resulting from lipoperoxidative events, were suggested. Furthermore, attention was focused on techniques aimed to ascertain free radical-mediated damage on proteins or DNA (such as determination of carbonyl groups in proteins and hydroxynucleotides). The techniques for the detection of free radical intermediates (ESR of spin traps and low-level and induced chemiluminescence) were discussed. For large-scale studies, the determination of the global antioxidant capacity of plasma was proposed.

The third part of the round table was dedicated to a discussion of the use of antioxidants in prevention and/or therapy of neurological disorders. The discussion ended with an enumeration of various therapeutic or prophylactic applications of natural or synthetic antioxidants. The following problems were especially discussed:

– Is the recommended daily intake of natural antioxidants sufficient to prevent such diseases or is it advisable to increase this intake to quench enhanced generation of prooxidant species?

The question was raised as to whether we can improve on nature, for instance with so called SOD mimics or vitamin analogues. The question of whether supplementation of antioxidant vitamins is necessary or whether an appropriate nutrition would be sufficient for prevention of free radical-related diseases remained open.

– What is the strategy for using "emergency" antioxidants in contrast to antioxidants with prolonged action? What are the structural requirements for the antioxidants of these two types?

– To prevent free radical-mediated brain damage, are mixtures of various antioxidants more effective than the use of single ones? If antioxidant mixtures are advisable, is it possible to recommend a specific "cocktail?" No definitive answer to this important question could be given.

In the fourth and last part of the discussion, everybody agreed about the need for large epidemiological studies to ascertain the efficiency of antioxidant supplementation. Dr. L. Prilipko, as a representative of WHO, emphasized in his concluding remarks the importance of large scale studies that would allow consensus statements for public health authorities. Studies similar to the WHO/Monica project would be desirable also for brain ischemia and mental disorders. Cooperation between the WHO and scientific societies could help to develop a network of research, and first contacts have now been established in this direction with the Society for Free Radical Research.

References

Bakalova RA, Nekrasov AS, Lankin VZ, Kagan VE, Stoytchev TS, Evstigneeva RP (1988) A mechanism of the inhibitory effect of alpha-tocopherol and its synthetic analogs on the oxidation of linoleic acid catalyzed by lipoxygenase from reticulocytes. Proc. Natl. Acad. Sci. USSR 299 (4): 1008–1001

Boscoboinik B, Szweczyk A, Hensey CE, Azzi A (1991) Inhibition of cell proliferation by α-tocopherol: role of protein kinase C. J. Biol. Chem, in press

Carney JM, Starke-Reed PE, Oliver CN, Landum RW, Cheng MS, Wu JF, Floyd R.A (1991) Reversal of age-related increase in brain protein oxidation, decrease in enzyme activity, and loss in temporal and spatial memory by chronic administration of the spin-trapping compound N-*tert*-butyl-α-phenylnitrone. *Proc. Natl. Acad. Sci.*, USA 88: 3633–3636

Clemens MR, Ladner C, Ehninger G, Einsele H, Renn W, Buhler E, Waller HD, Gey KF (1990) Plasma vitamin E and beta-carotene concentrations during radiochemotherapy preceding bone marrow transplantation. Am. J. Clin. Nutr. 51(2): 216–219

Douglas CE, Chan AC, Choy PC (1986) Vitamin E inhibits platelet phospholipase A2. Biochim. Biophys. Acta 876(3): 639–645

Elkashef AM, Ruskin PE, Bacher N, Barrett D (1990) Vitamin E in the treatment of tardive dyskinesia. Am. J. Psych., 147(4): 505–6

Elson CE (1991) The impact of palm oil on experimental carcinogenesis In: Packer L, Fuchs J (eds) *Vitamin E: Biochemistry and Clinical Applications*. Marcel Dekker, Inc., New York, NY, in press

Floyd R (1990) Role of oxygen free radicals in carcinogenesis and brain ischemia. *FASEB J* 4: 2587–2597

Gey KF, Puska P, Jordan P, Moser UK (1991) Inverse correlation between plasma Vitamin E and Mortality from ischemic heart disease in cross-cultural epidemiology. *Am. J. Clin. Nutr.* 53: 1–9

Gutteridge JMC, Halliwell B (1990) Reoxygenation injury and antioxidant protection: a tale of two paradoxed. Arch. Biochem. Biophys., 283(2): 223–226

Haberal M, Hamaloglu E, Bora S, Oner G, Bilgin N (1988) The effects of vitamin E on immune regulation after thermal injury. Burns Incl Therm. Inj. 14(5): 388–393

Halevy O, Sklan D (1981) Inhibition of arachidonic acid oxidation by beta-carotene, retinol, and alpha-tocopherol. Biochim. Biophys. Acta. 918(3): 304–307

Halliwell B (1989) Oxidants and the central nervous system: some fundamental questions. Is oxidant damage relevant to Parkinson's disease, Alzheimer's disease, traumatic injury or stroke? Acta. Neurol. Scand., 126: 23–33

Hiramatsu M, Packer L (1991) Interactions of water- and lipid-soluble antioxidants with alpha-tocopheroxyl radicals in membranes. In: Ozawa T (ed) New trends in biological chemistry, Japan Sci. Soc. Press, Tokyo/Springer-Verlag, Berlin pp. 323–331

Hiramatsu M, Velasco RD, Packer L (1990) Vitamin E radical reaction with antioxidants in rat liver. Free Rad. Biol. Med. 9: 459–464

Jesberger JA, Richardson JS (1991) Review of oxygen free radicals and brain dysfunction. Intern. J. Neurosci 57: 1–17

Kagan V, Serbinova E, Packer L (1990a) Antioxidant effects of ubiquinones in microsomes and mitochondria are mediated by tocopherol recycling. Biochem. Biophys. Res. Commun, 169(3): 851–857

Kagan VE, Serbinova EA, Packer L (1990b) Generation and recycling of radicals from phenolic antioxidants. Arch. Biochem. Biophys., 280(1): 33–39

Kato A, Yamaoka M, Tanaka A, Komiyama K, Umezawa I (1985) Physiological effect of tocotrienol. J Jpn Oil Chem Soc 34: 375–376

Maguire JJ, Wilson DS, Packer L (1989) Mitochondrial electron transport linked tocopheroxyl radical reduction. J. Biol. Chem. 264(36): 21462–21465

Mahoney CV, Azzi A (1988) Vitamin E inhibits protein kinase C activity. Biochem. Biophys. Res. Commun 154(2): 694–697

Mehlhorn RJ, Sumida S, Packer L (1989) Tocopheroxyl radical persistence and tocopherol consumption in liposomes and in vitamin E-enriched rat liver mitochondria and microsomes. J. Biol. Chem., 264(23): 13448–13452

Milei J, Boveris A, Llesuy S, Molina HA, Storino R, Ortega D, Milei SE (1986) Amelioration of adriamycin-induced cardiotoxicity in rabbits prenylamine and vitamins A and E. Am. Heart J. 111(1): 95–102

Nordmann R (1987) Oxidative stress from alcohol in the brain. In: Lindros KO, Yilkahrl R, Kiianmaa K (eds) Advances in biomedical alcohol research. Pergamon Press, Oxford, New York, pp 75–82

Nordmann R, Ribiere C, Rouach H (1990) Ethanol-induced lipid peroxidation and oxidative stress in extrahepatic tissues. Alcohol Alcoholism 25: 231–237

Ogunmekan AO, Hwang PA (1989) Randomized, double-blind, placebo-controlled, clinical trial of d-alpha-tocopheryl acetate (Vitamin E), as add-on therapy, for epilepsy in children. Epilepsia 30(1): 84–89

Packer L (1991) Membrane stabilization by antioxidant recycling. In: Fernandez-Gomez, J (ed) *Advances in biomembranes and biomaterials.* Birkhauser Verlag, pp 140–154

Packer L, Landvik S (1989) Vitamin E. Introduction to its biochemistry and health benefits. In: Diplock AT, Machlin LJ, Packer L, Pryor WA (eds) Vitamin E biochemistry and health implications. New York Academy of Sciences 570, pp 1–6

Packer L, Maguire JJ, Mehlhorn RJ, Serbinova E, Kagan VE (1989) Mitochondria and microsomal membranes have a free radical reductase activity that prevents chromanoxyl radical accumulation. Biochem. Biophys. Res. Commun 159(1): 229–235

Packer L, Serbinova EA, Khwaja S, Reznick AZ, Kagan VE (1991) Myocardial ischemia reperfusion injury: mechanisms of interventions and protection by antioxidants-effects of palm oil vitamin E. In: Oon Chong Hau (ed.) Cardiac rehabilitation and preventive cardiology. Annals of Academy & Medicine, Singapore, in press

Qureshi AA, Burger WC, Peterson DA, Elson CE (1986) The structure of an inhibitor of cholesterol biosynthesis isolated from barley. J Biol Chem 261: 10544–10550

Reddanna P, Rao MK, Reddy CC (1985) Inhibition by 5-lipoxygenase by vitamin E FEBS Lett., 193: 39–43

Richard C, Lemonnier F, Thibault M, Couturier M, Auzepy P (1990) Vitamin E deficiency and lipoperoxidation during adult respiratory distress syndrome. Crit. Care Med. 18(1): 4–9

Riemersma RA, Wood DA, Macintyre CCA, Elton RA, Gey FA, Oliver MF (1991) Risk of angina pectoris and plasma concentrations of vitamins A, C, and E and carotene. *The Lancet* 337

Rouach H, Ribiere C, Park MK, Saffar C, Nordmann R (1981) Lipid peroxidation and brain mitochondrial damage induced by ethanol. Bioelectrochem. Bioenergetics, 18: 211–217

Serbinova E, Kagan V, Han D, Packer L (1991a) Free radical recycling and intramembrane mobility in the antioxidant properties of alpha-tocopherol and alpha-tocotrienol. Free Rad. Biol. Med., 10: 263–275

Serbinova EA, Kagan VE, Han D, Packer L (1991b) d-Alpha-tocotrienol is a more powerful membrane antioxidant than d-alpha-tocopherol. In: Davies KJA (ed) Oxidative damage and repair. Pergamon Press, in press

van Kuijk FJGM, Sevanian A, Handelman GJ, Dratz EA (1987) A new role for phospholipase A2: protection of membranes from lipid peroxidation damage. Trends Biochem Sci 12: 31–34

Reactive Oxygen Species and the Central Nervous System

B. Halliwell

Summary

Radicals are species containing one or more unpaired electrons. The oxygen radical superoxide (O_2^-) and the non-radical oxidants hydrogen peroxide (H_2O_2) and hypochlorous acid (HOCl) are produced during normal metabolism and perform several useful functions. Excessive production of O_2^- and H_2O_2 can result in tissue damage, which often involves generation of highly reactive hydroxyl radical (·OH) and other oxidants in the presence of "catalytic" iron or copper ions. A major form of antioxidant defence is the storage and transport of iron or copper ions in forms that will not catalyze formation of reactive radicals. Tissue injury, e.g., by ischaemia or trauma, can cause increased iron availability and accelerate free radical reactions. This may be especially important in the brain, since areas of this organ are rich in iron and cerebrospinal fluid cannot bind released iron ions. Oxidative stress upon nervous tissue can produce damage by several interacting mechanisms, including rises in intracellular free Ca^{2+} and, possibly, release of excitatory amino acids. Recent suggestions that free radical reactions are involved in the neurotoxicity of aluminium and in damage to the substantia nigra in Parkinson's disease are reviewed. Finally, the nature of antioxidants is discussed, with a suggestion that antioxidant enzymes and chelators of iron ions may be more generally useful protective agents than chain-breaking antioxidants. Careful precautions must be taken in the design of antioxidants for therapeutic use.

Introduction

Although aerobes need oxygen (O_2) for survival, O_2 concentrations greater than those present in normal air have long been known to cause damage (Balentine 1982). The signs of O_2 toxicity depend upon the organism under study, its age, physiological state, and diet. For example, pure O_2 is less toxic to adult humans than to adult rats, and less toxic to newborn rats than to adult rats (reviewed in Halliwell and Gutteridge 1989). High pressure O_2 causes *acute CNS toxicity in humans,* leading to convulsions. This has been a problem in diving and must be taken into account in using hyperbaric oxygen therapy, e.g., in the treatment of gas gangrene, multiple sclerosis and in combination with radiotherapy (O_2 aggravates the damage done by ionizing radiation to cells).

The acute effects of hyperbaric O_2 have often been attributed to direct inactivation of enzymes (Balentine 1982), although the evidence for this is not convincing. However, the slower-acting toxic effects of elevated O_2 have often been suggested to involve "oxygen radicals" (Balentine 1982; Halliwell and Gutteridge 1989; Fridovich 1978). Such radicals may also be involved in the acute effects produced by high-pressure O_2 (Turrens et al. 1984).

The Nature of Radicals

Electrons within atoms and molecules occupy regions of space known as orbitals. Each orbital can hold a maximum of two electrons. For example, the two electrons that form a covalent bond occupy the same (molecular) orbital but have opposite spins. If an orbital contains only one electron, that electron is said to be *unpaired.* A free radical is defined as any species capable of independent existence (hence the term "free") that contains one or more unpaired electrons. This broad definition encompasses a wide range of species, some of which are listed in Table 1.

The diatomic oxygen molecule qualifies as a radical since it has two unpaired electrons, each located in a different orbital but both with the same spin quantum number. This parallel spin is one reason for the poor reactivity of O_2, despite its powerful oxidizing nature. According to thermodynamics, the complex organic compounds of the human body, almost entirely non-radical species, should immediately combust in the O_2 of the air. However, if O_2 attempts to oxidize a molecule directly by accepting a pair of electrons from it, both of these electrons must have spin opposite to that of the unpaired electrons in the O_2 so as to fit into the vacant spaces in the O_2 orbitals. A pair of electrons

Table 1. Some biologically relevant free radicals

Type of radical	Examples	Comments
Hydrogen-centred	H atom (1 proton, 1 electron)	H atom abstraction from other atoms often initiates radical chain reactions, e.g., $\cdot OH$ can initiate lipid peroxidation by abstracting H from the fatty acid side-chains of membrane lipids: $L-H+{}^{\bullet}OH \rightarrow L^{\bullet}+H_2O$.
Carbon-centred	Trichloromethyl radical, $CCl_3^{\bullet}$	Major agent in carbon tetrachloride hepatotoxicity.
Sulphur-centred	Thiyl radicals, $R\text{-}S^{\bullet}$	Reactive radicals produced during oxidation of thiol compounds (accelerated by transition metals). $RS^{\bullet}$ may react with O_2 to give $RSO^{\bullet}$ and $RSO_2^{\bullet}$ species.
Nitrogen-centred	Phenyldiazine radical, $C_6H_5N{=}N^{\bullet}$	Involved in phenylhydrazine toxicity to erythrocytes.

▲O_2▲ — The O_2 molecule has two unpaired electrons with parallel spin

▲ ▼ — In a covalent bond, both electrons have opposite spin

▲O_2▲ + ▲ ▼ SLOW — The requirement for spin inversion before two electrons can be accepted slows down the reaction of O_2 with non-radicals (most biomolecules)

O_2 —— le ——▶ [▲ ▼ O_2 ▲]$^-$
superoxide
1 unpaired electron

O_2^- —— 2H, le ——▶ H▲ ▼ O_2▼ ▲H
hydrogen peroxide
no unpaired electrons, not a radical

Fig. 1. The spin restriction of oxygen, shown diagrammatically

in the same orbital (e.g., a covalent bond) would not meet this criterion, since they have spins opposite to each other. This *spin restriction* slows down the reaction of O_2 with non-radicals (most biological molecules; Fig. 1). Oxygen much prefers to react with radicals (Table 1), accepting its electrons one at a time. Thus combination of carbon-centred or sulphur-centred radicals with O_2 is often very fast (Table 1).

Over 90 % of the O_2 taken up by the human body is used by mitochondrial cytochrome oxidase, which adds four electrons onto each O_2 molecule to generate two molecules of water. This enzyme, like most others that use O_2, has transition metal ions at its active sites. Ions of transition metals such as iron, vanadium, copper or titanium have variable valencies and can transfer single electrons. Indeed, reduction of O_2 to $2H_2O$ by cytochrome oxidase proceeds in a stepwise fashion, with various partially reduced forms of oxygen held firmly bound to the enzyme and not released into free solution. Thus cytochrome oxidase does not release reactive oxygen radicals into its surroundings.

Several non-radical compounds "autoxidize" on exposure to air; examples are adrenalin, noradrenalin, Dopa, ascorbic acid, 6-hydroxydopamine and thiols such as GSH. The rates of these "autoxidations" depend on the amount of contaminating metal ions in the reagents (Table 2); it may be that the "autoxidizable" compounds would not oxidize at all if metal ion contamination could be removed completely (an impossible task!). For example, manganese ions accelerate the oxidation of catecholamines to produce quinones, semiquinones and oxygen radicals; this has been suggested to account for the degeneration of catecholamine neurones that has been reported in miners of manganese-containing ores (Halliwell 1984; Archibald and Tyrce 1987; Liccione and Maines 1988), the so-called "locura manganica." The nervous system is rich in oxidizable catecholamines and mixtures of Fe^{3+} or Cu^{2+} ions with Dopa or Dopamine have been reported to stimulate peroxidation of lipids (Sotomatsu et al. 1990).

Table 2. Role of metal ions in converting less reactive to more reactive species

Less reactive species	Metal ion Fe/Cu[a]	More reactive species
O_2^- H_2O_2	Fe/Cu[a]	•OH
Lipid peroxides (ROOH)	Fe/Cu[b]	RO• (alkoxyl), $RO_2^•$ (peroxyl), cytotoxic aldehydes
Thiols (RSH)	Fe/Cu plus O_2[c]	O_2^-, H_2O_2, thiyl (RS•), •OH
NAD(P)H	Fe/Cu plus O_2[c]	NAD(P)•, O_2^-, H_2O_2, •OH
Ascorbic acid	Cu/Fe[d]	Semidehydroascorbate radical, •OH, H_2O_2, degradation products of ascorbate
Catecholamines, related autoxidizable molecules	Fe/Cu/Mn plus O_2[c]	O_2^-, H_2O_2, •OH, semiquinones (or other radicals derived from the oxidizing compounds)

[a] The iron or copper catalyzed Haber-Weiss reaction
[b] Lipid peroxide decomposition is metal ion-dependent and produces highly cytotoxic products such as 4-hydroxy-2,3-*trans*-nonenal.
[c] Most so-called autoxidations are stimulated by traces of transition metal ions and proceed by free radical mechanisms.
[d] Copper ions are especially effective in decomposing ascorbic acid, and ascorbate/copper or ascorbate/iron mixtures are cytotoxic.

Superoxide Radical: A Useful Species?

Acceptance of a single electron by the O_2 molecule forms the superoxide radical, O_2^-, with one unpaired electron (Fig. 1). The discovery, by McCord and Fridovich, of enzymes that appear to have evolved to scavenge specifically O_2^- in living cells *(superoxide dismutases)* led to the proposal that O_2^- is a major agent responsible for O_2 toxicity and that superoxide dismutases are important antioxidant defences. This proposal is supported by a wide range of evidence (Fridovich 1978), including results obtained using the techniques of modern molecular biology (Van Loon et al. 1986; Touati 1989; Chan et al. 1990).

Superoxide is formed in vivo in a variety of ways. A major source is the activity of electron transport chains in mitochondria and endoplasmic reticulum. Some of the electrons passing through these chains "leak" directly from reduced electron transfer carriers onto O_2. Since O_2 accepts electrons one at a time, O_2^- is formed. The rate of leakage at physiological O_2 concentrations is <5 % of total electron flow through the chains, but it rises as O_2 concentration is increased (Halliwell and Gutteridge 1989). In addition, activated phagocytic cells produce O_2^-, which plays an important part in the mechanism by which engulfed bacteria are killed (Curnutte and Babior 1987). Phagocytes able to produce O_2^- include monocytes, neutrophils, eosinophils and macrophages of various types (Curnutte and Babior 1987), including the microglial cells of the brain (Colton and Gilbert 1987). It has been suggested that microglial O_2^- generation could serve as a protection against infective organisms in the CNS (Colton and Gilbert 1987).

The evidence supporting the superoxide theory of O_2 toxicity is substantial (Halliwell and Gutteridge 1989; Fridovich 1978; Van Loon et al. 1986; Touati 1989; Chan et al. 1990), but the exact mechanism by which the putative excess O_2^- generation at increased O_2 levels exerts toxic effects is not completey clear. Indeed, O_2^- itself is a poorly reactive radical species, although it can inactivate a few enzymes directly, such as creatine kinase (McCord and Russell 1988), and some iron-sulphur proteins, such as bacterial aconitase (Gardner and Fridovich, 1991).

However, the number of cellular targets known to be attached directly by O_2^- in animals is small. Indeed, under certain circumstances, O_2^- production is a useful process, e.g., in the bacterial killing mechanisms of phagocytes (Curnutte and Babior 1987). Evidence is also accumulating that several cell types, such as fibroblasts, lymphocytes and vascular endothelial cells, produce and release small amounts of O_2^- (Murrell et al. 1990; Maly 1990; Meier et al. 1990).

Another free radical released by vascular endothelial cells and phagocytes such as neutrophils and macrophages is nitric oxide (NO), which is thought to be identical or closely related to the endothelium-derived relaxing factor (EDRF). NO reacts with O_2^- at physiological pH to yield a non-radical product, peroxynitrite (Saran et al. 1990). There is considerable debate in the literature as to whether this interaction of O_2^- and NO might be damaging to cells (Radi et al. 1991) or whether it may represent a physiological process affecting vascular muscle tone (Halliwell 1989 a). In any case, O_2^- often has a vasoconstrictor effect in biological systems (Laurindo et al. 1991).

Reaction of O_2^- and NO might conceivably be relevant in the brain, since nitric oxide synthetase appears to be widely distributed in brain tissue (Hope et al. 1991).

Hydrogen Peroxide

Superoxide dismutases remove O_2^- by converting it into hydrogen peroxide.

$$2O_2^- + 2H^+ \rightarrow H_2O_2 + O_2 \tag{1}$$

Several other enzymes that produce H_2O_2 exist in human tissues, such as L-amino acid oxidase, glycolate oxidase and monoamine oxidase. Indeed, oxidative deamination of dopamine by monoamine oxidase is the main catabolic pathway for dopamine within dopamine nerve terminals (Cohen 1988). Cohen (1988) has suggested that an accelerated turnover of dopamine in Parkinson's disease, leading to increased H_2O_2 formation, may provoke an oxidative stress (i.e., an increase in the generation of oxygen-derived species beyond the ability of antioxidant defences to cope with them; Sies 1991) within surviving dopamine terminals and accelerate their destruction. Another possibility is that the side effects of prolonged therapy with L-Dopa in this disease might relate to increased H_2O_2 formation.

H_2O_2 is an oxidizing agent, although not very reactive. Unlike O_2^-, however, H_2O_2 crosses cell membranes easily. H_2O_2 does not qualify as a radical because it contains no unpaired electrons (Fig. 1), and it can be removed within human cells by the action of two enzymes, catalase and selenium-dependent glutathione peroxidase (Halliwell and Gutteridge 1989; Sies 1991).

Hydroxyl Radical and Related Species

H_2O_2 is known to be toxic to many cells, but toxicity is not usually mediated by a direct effect of the H_2O_2, except at high concentrations. Instead, the H_2O_2 forms highly-oxidizing radicals. It combines with Fe^{2+} ions to form *hydroxyl radical,* ·OH, by the *Fenton reaction.*

$$Fe^{2+} + H_2O_2 \rightarrow Fe^{3+} + \cdot OH + OH^- \quad (2)$$

It is likely that the initial product of reaction of Fe^{2+} and H_2O_2 is another highly-oxidizing species, an iron-oxygen complex called *ferryl,* that then decomposes to yield ·OH (Halliwell and Gutteridge 1990 a). Copper ions also react with H_2O_2 to form ·OH (Halliwell and Gutteridge 1990 a; Aruoma et al. 1991).

Hydroxyl radical reacts at great speed with almost every molecule found in living cells, including DNA (causing strand breakage and chemical alterations to the purine and pyrimidine bases), membrane lipids and carbohydrates. Indeed, the cytotoxic action of H_2O_2 upon most, if not all, mammalian cells seems to involve DNA damage. Depending on the cell-type studied, some or all of this DNA damage may be mediated by reaction of H_2O_2 with iron and/or copper ions, bound at or close to the DNA, to form ·OH (reviewed in Halliwell 1987; Halliwell and Aruoma 1991).

Another potentially devastating effect of ·OH is its action upon membrane lipids (Halliwell and Gutteridge 1989). Various reactive species formed in mixtures of O_2^-, H_2O_2 and iron or copper ions, including ·OH, initiate the process of *lipid peroxidation* by abstracting a hydrogen atom from a polyunsaturated fatty-acid side chain (written lipod-H below) in a membrane phospholipid. Abstraction of a hydrogen atom leaves behind a carbon-centred radical (lipid·) in the membrane (Table 1). The most likely fate of this under aerobic conditions is reaction with O_2 to form a peroxyl radical (often shortened to "peroxy" radical).

$$\text{lipid}\cdot + O_2 \rightarrow \text{lipid-}O_2\cdot \quad (3)$$

Peroxyl radicals can abstract hydrogen atoms from adjacent fatty acid side-chains to continue the chain reaction.

$$\text{lipid-}O_2\cdot + \text{lipid-H} \rightarrow \text{lipid-}O_2H + \text{lipid}\cdot \quad (4)$$

Thus abstraction of a single hydrogen can conceivably lead to conversion of many membrane lipids into *lipid hydroperoxides* (lipid-O_2H). The existence of these within a membrane severely disrupts its functioning, decreasing fluidity and allowing ions such as Ca^{2+} to leak across the membrane (Halliwell and Gutteridge 1989). Iron and copper ions can contribute to lipid peroxidation by causing formation of initiating species. They can also stimulate peroxidation by reacting with lipid hydroperoxides and decomposing them to peroxyl radicals and alkoxyl radicals (lipid-O·), which can cause further peroxidation (Halliwell and Gutteridge 1989, 1990 a). Products of these complex decomposition reactions include hydrocarbon gases and a wide range of toxic aldehydes. Of these aldehydes, most attention in the literature is usually devoted to malonaldehyde (sometimes called malondialdehyde or MDA), but this is fairly harmless in comparison with such highly cytotoxic products as 4-hydroxy-2,3-*trans*-nonenal (Esterbauer 1985). These aldehydes can damage adjacent cells; membrane-bound enzymes and receptors can also be inactivated. For example, the binding of serotonin to its receptors in rat cortex membranes (Muakkassah-Kelly et al. 1982) and the binding of spiroperidol to receptors in human caudate and putamen (Andorn et al. 1988) are decreased by peroxidation of the membranes.

Formation of ·OH and other reactive species usually requires a reaction between reduced iron (Fe^{2+}) or copper (Cu^+) ions and H_2O_2 (Eq. 2). The Fe^{3+} or Cu^{2+} generated can be re-reduced by ascorbic acid or by O_2^-. Thus, under certain circumstances, both O_2^- and ascorbic acid can accelerate the copper- or iron-dependent formation of ·OH from H_2O_2 (Halliwell and Gutteridge 1990 a; Aruoma et al. 1991). Reaction of Fe^{3+} with O_2^- proceeds with intermediate formation of *perferryl*, another oxidizing species but apparently a much weaker oxidant than ferryl or ·OH at physiological pH (Halliwell and Gutteridge 1990 a).

$$Fe^{3+} + O_2^- \rightleftarrows (Fe^{3+} - O_2^- \leftrightarrow Fe^{2+} - O_2) \rightleftarrows Fe_{2+} + O_2 \text{ perferryl} \tag{5}$$

Combining this equation with Eq. 2, we arrive at the overall reaction that has been called the *iron-catalyzed Haber-Weiss reaction*

$$O_2^- + H_2O_2 \xrightarrow[\text{Catalyst}]{\text{Fe}} \cdot OH + OH^- + O_2 \tag{6}$$

Copper ions also catalyze this overall reaction (Halliwell and Gutteridge 1990 a; Aruoma et al. 1991).

Equation 6 implies that much of the toxicity of O_2^- and H_2O_2 in vivo is due to their reaction with iron or copper ions to form ·OH and other oxidizing species. Considerable evidence exists to support this suggestion (Halliwell and Gutteridge 1989, 1990 a; Fridovich 1978; Halliwell 1987; Aruoma et al. 1991; Kyle et al. 1988; Imlay and Linn 1988; Meneghini 1988), although reaction 6 is not likely to be the only mechanism of cell injury by excess O_2^-. It follows that *the nature of the damage done to cells by H_2O_2 and O_2^- will be affected by the location of metal* *ions catalysts of reaction 6 within the cells.* It also follows that, *if no catalytic metal ions are available, then O_2^- and H_2O_2 will have limited, if any, damaging effect.* These principles underline the importance of examining the availability and distribution of "catalytic" metal ions in explaining oxidative damage to cells (Halliwell and Gutteridge 1986, 1990 a; Aruoma et al. 1991; Halliwell 1987). In general, iron and copper ions act to convert poorly reactive species (O_2^-, H_2O_2, thiols, lipid-O_2H) into more reactive, cytotoxic ones ($\cdot OH$, ferryl, aldehydes etc.) as summarized in Table 2.

The Source of Iron

Iron is a remarkably useful metal in nature; we depend on it to transport (haemoglobin), store (myoglobin), and utilize (cytochromes, cytochrome oxidase) oxygen for respiration. It is also an essential component in the active sites of many enzymes (e.g., aconitase, proline hydroxylase), including enzymes that protect against oxidants (iron-containing superoxide dismutase enzymes in bacteria and some plants, catalase in most aerobic organisms). In mammals, iron is absorbed from the gut and enters the plasma attached to the protein transferrin, which binds two moles of Fe^{3+} per mole of protein with very high affinity at pH 7.4. In normal human plasma, the average iron loading of transferrin is 20 % – 30 % of maximum. Hence there is an excess of iron-binding capacity, which means that the content of "free" ionic iron in plasma should be effectively nil, a result confirmed by experiments (Halliwell and Gutteridge 1986).

Transferrin enters cells by endocytosis, and the pH of the vacuole containing it is then lowered. This facilitates release of Fe^{3+} from the protein. The unloaded transferrin (apotransferrin) is ejected from the cell, and the iron ions released from it are used for the synthesis of intracellular iron proteins. Excess iron is stored in the protein ferritin. At present, the chemical nature and subcellular distribution of the "non-protein-bound iron pool" within cells are completely unknown, except that the "pool" seems to be kept very small.

Iron ions attached to the transport protein transferrin, the neutrophil-derived protein lactoferrin, the iron-storage proteins ferritin and haemosiderin, the O_2-binding proteins haemoglobin, myoglobin or leghaemoglobin or to other iron proteins (e.g., ferredoxin) are thought to be *incapable* of reacting with H_2O_2 to form $\cdot OH$ (Halliwell and Gutteridge 1986; 1990 a; Puppo and Halliwell 1989). Lactoferrin and transferrin are resistant to damage by O_2^-, H_2O_2 and other oxidants, which makes these proteins a good form of transport for iron (Halliwell and Gutteridge 1986; Halliwell et al. 1988 b). Thus iron bound to these proteins will not stimulate lipid peroxidation, $\cdot OH$ formation or "autoxidation" reactions. By contrast, small amounts of Fe^{2+} can be reductively mobilized from ferritin by O_2^- and by radicals generated during lipid peroxidation (reviewed in Halliwell and Gutteridge 1990 a; Bolann and Ulvik, 1990). Hence catalytic iron is available within cells, as a mobile "pool" and as iron releasable from ferritin, and so O_2^- and H_2O_2 must be scavenged by intracellular enzymes before they can come into contact with this available iron.

By contrast, in blood plasma from healthy humans, no iron is available to stimulate lipid peroxidation or ·OH radical formation (Halliwell and Gutteridge 1986, 1990 a). Plasma contains no catalase and little GSH, superoxide dismutase or glutathione peroxidase. Gutteridge and I have argued (Halliwell and Gutteridge 1986, 1989, 1990 a, b) that O_2^- and H_2O_2 generated extracellularly may be employed for *useful* purposes, such as cellular signalling and inactivation of EDRF. They are not dangerous because, in the absence of catalytic iron ions, they cannot lead to formation of ·OH and other highly oxidizing species. Thus *a major form of antioxidant defence in human plasma is the prevention of metal ions from participating in the generation of reactive radical species.* Copper ions are similarly unavailable in human plasma; most or all of plasma copper is attached to the protein caeruloplasmin, which has antioxidant properties (Halliwell and Gutteridge 1986, 1990 a, b).

Haemoglobin – A Dangerous Protein

Haemoglobin is transported in erythrocytes, which are rich in the antioxidant defence enzymes catalase, superoxide dismutase and glutathione peroxidase. However, pure haemoglobin (and myoglobin) can be degraded by excess H_2O_2 to release "catalytic" iron ions from the haem ring (Gutteridge 1986; Puppo and Halliwell 1988). In addition, haemoglobin reacts with H_2O_2 to form an oxidizing species capable of stimulating lipid peroxidation (Kanner et al. 1987, Puppo and Halliwell 1988). The nature of this oxidizing species is not known. Reaction of H_2O_2 with the protein probably generates a haem ferryl species plus an amino acid radical. In the case of myoglobin, it has been suggested that a tyrosine peroxyl radical, capable of abstracting hydrogen and initiating lipid peroxidation, is formed upon exposure of the myoglobin to H_2O_2 (Davies 1990).

Thus, as Fig. 2 emphasizes, haemoglobin outside the erythrocyte is potentially a dangerous protein (Steele et al. 1991). Indeed, spasm of cerebral arteries can be a significant late complication of haemorrhagic stroke, and it has been proposed that release of haemoglobin from erythrocytes in the clot and subsequent free radical reactions are involved (Steele et al. 1991).

In general, however, the availability of "catalytic" iron or copper ions to stimulate free radical reactions in vivo is extremely limited and can often control the extent of tissue damage being done (e.g., O_2^- and H_2O_2 are unlikely to do serious damage unless such ions are present). However, *cell injury by almost any mechanism, including traumatic or ischaemic injury, has the potential to accelerate free radical reactions.* This is partly because injured and lysed cells release their intracellular iron pools into the surrounding environment (Halliwell and Gutteridge 1984, 1986; Halliwell et al. 1988 a), where the iron is then available to accelerate formation of damaging oxidant species. Bleeding as a result of injury, and subsequent haemoglobin liberation, may have a similar effect (Steele et al. 1991).

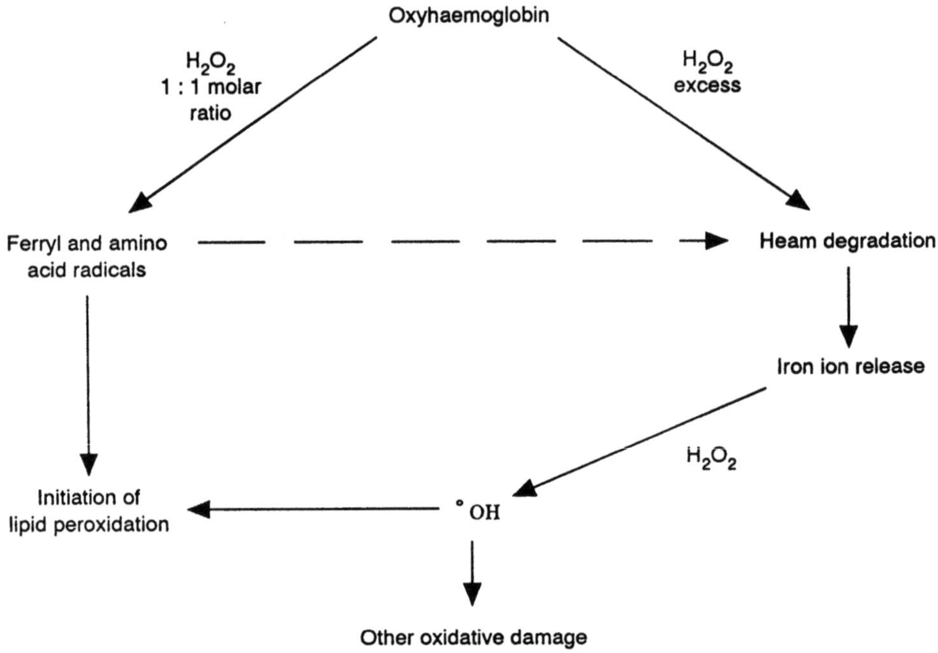

Fig. 2. Haemoglobin – a dangerous protein. Haemoglobin is normally transported inside erythrocytes, cells rich in antioxidant defence enzymes. Free haemoglobin is very sensitive to attack by H_2O_2, resulting in an exacerbation of oxidative damage

Oxidants, Iron, Copper and the Brain

Brain damage is a major clinical problem. Indeed, ischaemic or traumatic injuries to the brain or spinal cord are often said to result in more extensive tissue damage than do equivalent insults to other tissues. Free radical reactions have often been implicated in such damage (Steele et al. 1991; Demopoulos et al. 1982), particularly when ischaemic tissues are reoxygenated (Patt et al. 1988). Ischaemia appears to "prime" tissues (including brain) to respond upon oxygenation with increased rates of generation of O_2^- and H_2O_2 (Halliwell 1989 a; Patt et al. 1988; Kehrer 1989). These increased rates of oxidant generation in brain may sometimes be mediated by the enzyme xanthine oxidase in rat and gerbil tissues (Patt et al. 1988; Betz 1985), but the role of this enzyme is increasingly questioned (Kehrer 1989; Betz et al. 1991) and its significance in human brain is unclear at present. In addition, the importance of free radical generation upon reoxygenation as an injury mechanism may depend critically upon the extent and period of oxygen deprivation (Agardh et al. 1991).

The brain and nervous system may be especially prone to radical damage for a number of reasons. First, the membrane lipids are very rich in polyunsaturated fatty acid side chains. Second, the brain is poor in catalase activity and has only moderate amounts of superoxide dismutase and glutathione peroxidase (Cohen 1988), perhaps because intracellular O_2 tension in the normally functioning brain is low. Thirdly, areas of the human brain (e.g., globus pallidus and substantia nigra) are rich in iron, yet cerebrospinal fluid has no significant iron-binding capacity because its content of transferrin is very low (Gruener et al. 1991). Most brain iron is protein-bound, but little is known of its molecular nature (Youdim 1988a). Iron seems to play essential roles in learning and memory, perhaps because iron ions are required for the correct binding of certain neurotransmitters to their receptors (Youdim 1988a).

Thus a high content of brain iron may be essential, but its presence means that any injury to brain cells may result in release of iron ions that can stimulate free radical reactions. Homogenates of brain peroxidize very rapidly in vitro; indeed, peroxidation of ox-brain homogenate has been used for a long time as an assay to measure antioxidant activity, and the peroxidation is inhibited by several iron-chelating agents (Stocks et al. 1974). Ischaemic or traumatic injury might be thought of as essentially causing a partial disruption of brain tissue ("partial homogenization").

In addition, there is a high concentration of ascorbic acid in the grey and white matter of the CNS. The choroid plexus has a specific active transport system that raises ascorbate concentrations in the cerebrospinal fluid (CSF) to about ten times the plasma level, and neural tissue cells have a second transport system that concentrates intracellular ascorbate even more (Spector and Eells 1984). Ascorbic acid in the absence of transition metal ions has antioxidant properties (Frei et al. 1989; Halliwell 1990). However, if catalytic iron were generated in the CNS as a result of injury, ascorbic acid might then stimulate ·OH generation within the brain and CSF. Indeed, injecting aqueous solutions of iron salts or haemoglobin into the cortex of rats has been shown to cause transient focal epileptiform discharges, lipid peroxidation and persistent behavioural and electrical abnormalities (e.g., Rosen and Frumin 1979; Willmore at al. 1980, 1986).

The brain also contains copper, but almost nothing is known of its molecular nature or whether it ever becomes available to stimulate free radical reactions (Heiskala et al. 1988).

Oxidative Stress: The Molecular Targets

Let us suppose that ischaemia/reoxygenation injury or traumatic damage to the brain produces liberation of catalytic metal ions and an increase in the formation of reactive radicals. How then could further cellular damage occur? Lipid peroxidation is one possibility, and several groups have found some evidence for increased peroxidation after ischaemia (Watson and Ginsberg 1988; Michel et al. 1987) trauma (Hall and Braughler 1988a) iron injection (Willmore et al. 1980) or treatment with methyl mercury (Sarafian and Verity 1991).

It must be emphasised, however, that *damage caused by oxidative stress need not necessarily involve lipid peroxidation* (Halliwell 1987). Early events in human cells subjected to oxidative stress include DNA damage and consequent activation of poly (ADP-ribose) synthetase (this enzyme polymerises ADP-ribose residues derived from NAD^+ and can lead to depletion of cellular NAD^+ content), falls in ATP content, and rises in intracellular free Ca^{2+}, with consequent activation of Ca^{2+}-stimulated proteases that can cause such phenomena as "bleb" formation on the plasma membrane of the cells (Halliwell 1987; McConkey et al. 1988; Michel et al. 1987; Hall and Braughler 1988a). Figure 3 indicates the complexity of the changes that can occur in mammalian cells subjected to oxidative stress: it is very difficult to disentangle the sequence of events. Rises in intracellular free Ca^{2+} might be particularly damaging to neurones (Siesjö 1990), since several neurotoxins are thought to act in this way (Siesjö 1990; Komulainen and Bondy 1988).

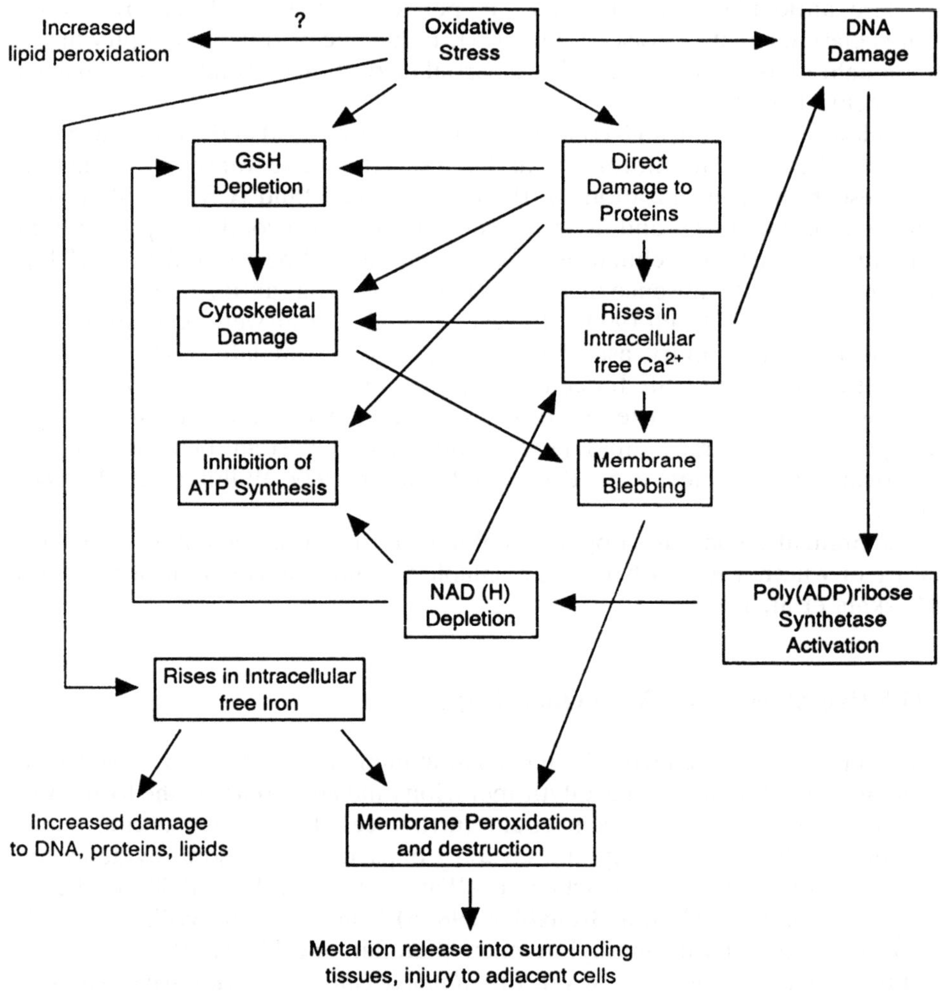

Fig. 3. Inter-relationship of oxidative stress mechanisms

An additional mechanism that has been suggested to account for the damage produced by ischaemia/reoxygenation in brain is the generation within the tissue of excitatory amino acids such as glutamate (Komulainen and Bondy 1988; Meldrum 1985; Siesjö 1988). It is therefore interesting to note recent claims that excitatory amino acids can be released from rat hippocampal slices as a result of exposure to O_2^- and H_2O_2 (Pellegrini-Giampietro et al. 1988), and that the toxicity of the excitatory neurotoxin kainic acid to cerebellar neurones in culture may involve increased oxygen radical formation (Dykens et al. 1987). In addition, it has been proposed that radicals could inactivate brain glutamine synthetase, the major enzyme responsible for glutamate removal (Schor 1988), as well as damaging other proteins (Carney et al. 1991). The acidosis resulting from ischaemia could also contribute to injury (Siesjö et al. 1985; Link 1988). Hence the failure of some groups to find lipid peroxides in injured nervous tissue does *not* rule out the possibility of oxidative damage.

Oxidative Damage and Parkinson's Disease

There has been considerable interest in a possible role of neurotoxins in causing Parkinson's disease, especially since the identification of 1-methyl-4-phenyl-1, 2, 5, 6 tetrahydropyridine (MPTP) as the toxic agent responsible for the sudden onset of Parkinsonian symptoms in a group of heroin addicts in the USA (Langston et al. 1987; Singer et al. 1987). Early suggestions that MPTP and its metabolites kill nigral cells by increasing intracellular oxidant formation have not been substantiated (Martinovits et al. 1986; Frank et al. 1987), and it seems more likely that MPTP is oxidized to MPP^+, which then inhibits mitochondrial electron transport and causes severe metabolic derangements (Langston et al. 1987; Singer et al. 1987; Turski et al. 1991). Of course, it is possible that interference with mitochondrial metabolism itself creates an oxidative stress, which worsens the damage (Adams and Odunze 1991). MPTP inhibits mitochondrial electron transport at complex I, the NADH dehydrogenase system. A defect in the function of mitochondrial complex I has recently been discovered in the substantia nigra of Parkinsonian patients, raising the possibility that Parkinson's disease may be the result of exposure to an environmental toxin acting in a way similar to MPTP (Shapira et al. 1990).

It is also *possible* that iron-dependent free radical reactions contribute to nigral cell damage in Parkinson's disease. Thus Dexter et al. (1989) have reported increased nigral iron and decreased brain ferritin levels in Parkinsonian patients, and they speculate that increased iron-dependent lipid peroxidation could contribute to cell destruction. Pall et al. (1987) found no change in CSF iron in Parkinson's disease, but a rise in CSF copper was demonstrable in some patients. Riederer and colleagues (1989; Sofic et al. 1991) reported increased nigral iron and *increased* ferritin in Parkinsonian brains, as well as decreased GSH. Of course, it must be remembered that nigral cell injury (by any mechanism) could lead to iron release and more free radical reactions, even if the cause of the injury had nothing to do with free radicals. The question that remains is, "does any such secondary radical generation contribute significantly to the disease pathology?"

It must also be remembered that some drugs, such as chlorpromazine and Dopa, have metal ion-binding capacity and may be capable of transporting metal ions into and out of the brain (Blake et al. 1985; Weiner et al. 1977). Thus care must be taken in interpreting changes in the brains of patients who may have been treated with a wide range of drugs. However, the changes in Parkinsonian substantia nigra seem specific to iron (Hirsch et al. 1991), which perhaps argues that they are not nonspecific consequences of drug treatment.

Oxidative Damage and Alzheimer's Disease

Although it seems unlikely to the author that excessive exposure to aluminium is the cause of Alzheimer's disease, interest continues in elucidating the precise role played by aluminium in this disorder. Aluminium ions are known to be neurotoxic, aluminosilicates have been identified at the cores of senile plaques and aluminium has been found within neurones bearing neurofibrillary tangles (reviewed in Birchall and Chappell 1988). It may be that Alzheimer's disease leads to an impaired blood-brain barrier and thus increased accumulation of aluminium within the brain. Some or all of this aluminium may enter attached to transferrin, which binds Al^{3+} with high affinity. There have been some reports that iron deposition is increased within or close to plaques (Youdim 1988 b). Gutteridge et al. (1985) showed that aluminium ions alone cannot promote membrane lipid peroxidation, but that they accelerate Fe^{2+}-dependent peroxidation. Pb^{2+} ions also promote Fe^{2+}-dependent peroxidation (Quinlan et al. 1988). Aluminium ions may also sometimes stimulate production of oxidants by human neutrophils (Stankovic and Mitrovic 1991). Thus it is *conceivable* that coaccumulation of aluminium and iron (Youdim 1988 b) might lead to oxidant damage, but this has yet to be proved relevant to Alzheimer's disease.

Antioxidant Therapy

The term "antioxidant" is frequently taken to mean a chain-breaking antioxidant inhibitor of lipid peroxidation, such as α-tocopherol. However, we have already seen that oxidative damage can occur without lipid peroxidation, and so inhibitors of lipid peroxidation might be therapeutically unsuccessful even in conditions where radicals are important in producing tissue injury (Halliwell 1987; Halliwell and Gutteridge 1984, 1986; Stankovic and Mitrovic 1991). It must also be noted that some chain-breaking antioxidants are capable of reducing Fe^{3+} to Fe^{2+} and can *exacerbate* iron-dependent radical reactions in non-lipid systems. For example, several flavonoids that inhibit lipid peroxidation (Laughton et al. 1989) *aggravate* iron ion-dependent radical damage to DNA.

Another approach has been to use antioxidant enzymes, especially as recombinant human superoxide dismutase is now available. Some positive results have been reported; for example, superoxide dismutase was claimed to diminish reperfusion injury to the ischaemic spinal cord in dogs (Lim et al. 1986), and oedema induced by low brain temperatures in rats (Chan et al. 1987), but much more work is required. Ebselen, a low molecular mass agent that removes H_2O_2 by a gluathione peroxidase-like activity (Muller et al. 1984) and inhibits lipid peroxidation (Hayaishi and Slater 1986), is also available for evaluation.

Another approach is the use of iron chelators that prevent iron ions from participating in such reactions as ·OH production and lipid peroxidation. The prototype is desferrioxamine, which inhibits most iron-dependent radical reactions and has been used successfully to diminish oxidative damage in several animal model systems of human disease (Halliwell 1987, 1989 b; Halliwell and Gutteridge 1986). However, most antioxidants (including desferrioxamine) probably do not cross the blood-brain barrier. Thus several scientists have sought alternative chelating agents. For example, a series of 21-aminosteroids with antioxidant activity has been developed (Braughler et al. 1987). One of these, U74006F, has given interesting results in animal model systems of traumatic or ischaemia/reoxygenation injury to the brain or spinal cord. Thus U74006F was observed to minimize the effects of reperfusion injury upon the brain of cats (Hall and Yonkers 1988 b), to protect against haemorrhagic shock or post-traumatic spinal cord ischaemia in cats (Hall et al. 1988 a; Hall 1988) and to minimize neurological damage after head injury in mice (Hall et al. 1988 b). Some related compounds, such as U78517F, may also have protective effects in animal model systems (Hall et al. 1990).

Although developed as inhibitors of lipid peroxidation (Braughler et al. 1987), the 21-aminosteroids presumably also inhibit other free radical reactions, although this remains to be investigated. Thus their protective action in some animal model systems should not necessarily be taken to mean that they are acting by inhibiting iron-dependent lipid peroxidation.

Conclusions

Free radical reactions are a part of normal human metabolism (Halliwell and Gutteridge 1989). When produced in excess, radicals can cause tissue injury. However, tissue injury can itself cause more radical reactions, which may (or may not) contribute to a worsening of the injury. The careful use of a range of antioxidants, combined with new methods for measuring free radical generation in humans (Halliwell and Grootveld 1987; Halliwell et al. 1988 c, 1992) is at long last enabling evaluation of the exact contribution of free radical reactions to human disease.

References

Adams JD Jr, Odunze IN (1991) Biochemical mechanisms of 1-methyl-4-phenyl-1, 2, 3, 6-tetrahydropyridine toxicity. Could oxidative stress be involved in the brain? Biochem Pharmacol 41: 1099–1105

Agardh CD, Zhang H, Smith ML, Siesjö BL (1991) Free radical production and ischemic brain damage: influence of postischemic oxygen tension. Int J Dev Neurosci 9: 127–138

Andorn AC, Bacon BR, Nguyen-hunh AT, Paolato SJ, Stitts JA (1988) Guanyl nucleotide interactions with dopaminergic binding sites labelled by [^{3}H]spiroperidol in human caudate and putamen: guanyl nucleotides enhance ascorbate-induced lipid peroxidation and cause an apparent loss of high affinity binding site. Mol Pharmacol 33: 155–162

Archibald FS, Tyrce C (1987) Manganese poisoning and the attack of trivalent manganese upon catecholamines. Arch Biochem Biophys 256: 638–650

Aruoma OI, Halliwell B, Gajewski E, Dizdaroglu M (1991) Copper ion-dependent damage to the bases in DNA in the presence of hydrogen peroxide. Biochem J 273: 601–604

Balentine JD (1982) Pathology of oxygen toxicity. Academic, New York

Betz AL (1985) Identification of hypoxanthine transport and xanthine oxidase activity in brain capillaries. J Neurochem 44: 574–579

Betz AL, Randall J, Martz D (1991) Xanthine oxidase is not a major source of free radicals in focal cerebral ischemia. Am J Physiol 260: H563–568

Birchall JD, Chappell JS (1988) Aluminium, chemical physiology and Alzheimer's disease. Lancet ii: 1008–1010

Blake DR, Winyard P, Lunec J, Williams A, Good PA, Gutteridge JMC, Rowley D, Halliwell B, Hider RC (1985) Cerebral and ocular toxicity induced by desferrioxamine. QJ Med 219: 345–355

Bolann BJ, Ulvik RJ (1990) On the limited ability of superoxide to release iron from ferritin Eur J Biochem 193: 899–904

Braughler JM, Pregenzer JF, Chase RL, Duncan LA, Jacobsen EJ, McCall JM (1987) Novel 21-amino steroids as potent inhibitors of iron-dependent lipid peroxidation. J Biol Chem 262: 10438–10440

Carney JM, Starke-Reed PE, Oliver CN, Landum RW, Cheng MS, Wu JF, Floyd RA (1991) Reversal of age-related charges in brain protein oxidation, decrease in enzyme activity, and loss, in temporal and spatial memory by chronic administration of the spin trapping compound N-tert-butyl-α-phenylnitrone Proc Natl Acad Sci USA 88: 3633–3636

Chan PH, Longar S, Fishman RA (1987) Protective effects of liposome-entrapped superoxide dismutase on posttraumatic brain edema. Ann Neurol 21: 540–547

Chan PH, Chu L, Chen SF, Carlson EJ, Epstein CJ (1990) Reduced neurotoxicity in transgenic mice overexpressing human copper-zinc-superoxide dismutase. Stroke 21 (Suppl III): 80–82

Cochrane CJ, Schraufstatter IV, Hyslop P, Jackson J (1988) Cellular and biochemical events in oxidant injury. In: Halliwell B (ed) Oxygen radicals and tissue injury. FASEB, pp 49–54

Cohen G (1988) Oxygen radicals and Parkinson's disease. In Oxygen Radicals and Tissue Ingury (ed. Haliwell, B) FASEB, pp 130–135

Colton CA, Gilbert DL (1987) Production of superoxide anions by a CNS macrophage, the microglia. FEBS Lett 223: 284–288

Curnutte JT, Babior BM (1987) Chronic granulomatous disease. Adv Hum Genet 16: 229–297

Davies MJ (1990) Detection of myoglobin-derived radicals on reaction of metmyoglobin with hydrogen peroxide and other peroxidic compounds. Free Radic Res Commun 10: 361–370

Demopoulos HB, Flamm E, Selligman M, Pietronigro DD (1982) Oxygen free radicals in central nervous system ischemia and trauma. In: Autor AP (ed) Pathology of oxygen. Academic, New York, pp 127–155

Dexter DT, Wells FR, Lees AJ, Agid F, Agid Y, Jenner P, Marsen CD (1989) Increased nigral iron content and alterations in other metal ions occurring in brain in Parkinson's disease. J Neurochem 52: 1830–1836

Dykens JA, Stern A, Trenkner E (1987) Mechanism of kainate toxicity to cerebellar neurons in vitro is analogous to reperfusion tissue injury. J Neurochem 49: 1222–1228
Esterbauer H (1985) Lipid peroxidation products: formation, chemical properties and biological activities. In: Poli G, Cheeseman KH, Dianzani MV, Slater TF (eds) Free radicals in liver injury. IRL, pp 29–47
Frank DM, Arora PK, Blumer JL, Sayre LM (1987) Model study on the bioreduction of paraquat, MPP^+ and analogs. Evidence against a "redox cycling" mechanism in MPTP neurotoxicity. Biochem Biophys Res Commun 147: 1095–1104
Frei B, England L, Ames BN (1989) Ascorbate is an outstanding antioxidant in human blood plasma. Proc Natl Acad Sci USA 86: 6377–6381
Fridovich I (1978) The biology of oxygen radicals. Science 209: 875–877
Gardner, PR, Fridovich, I (1991) Superoxide sensitivity of the *Escherichia coli* aconitase. J. Biol. Chem. 266: 19328–19333
Gruener N, Gozlan O, Goldstein T, Davis J, Besner I, Iancu TC (1991) Iron, transferrin and ferritin in cerebrospinal fluid of children Clin Chem 37: 263–265
Gutteridge JMC (1986) Iron promoters of the Fenton reaction and lipid peroxidation can be released from haemoglobin by peroxides. FEBS Lett 201: 291–295
Gutteridge JMC, Quinlan GJ, Clark IA, Halliwell B (1985) Aluminium salts accelerate peroxidation of membrane lipids stimulated by iron salts. Biochim Biophys Acta 835: 441–447
Hall ED (1988) Effects of the 21-aminosteroid U74006F on posttraumatic spinal cord ischemia in cats. J Neurosurg 68: 462–465
Hall ED, Braughler JM (1988 a) The role of oxygen radical-induced lipid peroxidation in acute central nervous system trauma. Oxygen radicals and tissue injury. (Halliwell, B ed) FASEB, pp 92–98
Hall ED, Yonkers PA (1988 b) Attenuation of postischemic cerebral hypoperfusion by the 21-aminosteroid U74006F. Stroke 19: 340–344
Hall ED, Yonkers PA, McCall JM (1988 a) Attenuation of hemorrhagic shock by the non-glucocorticoid 21-aminosteroid U74006F. Eur J Pharmacol 147: 299–303
Hall ED, Yonkers PA, McCall JM, Braughler JM (1988 b) Effects of the 21-aminosteroid U74006F on experimental head injury in mice. J Neurosurg 68: 456–461
Hall ED, Pazara KE, Braughler JM, Lineman KL, Jacobsen EJ (1990) Nonsteroidal lazaroid U78517F in models of focal and global ischemia. Stroke 21 (Suppl III): 83–87
Halliwell B (1984) Manganese ions, oxidation reactions and the superoxide radical. Neurotoxicology 5: 113–118
Halliwell B (1987) Oxidants and human disease: some new concepts. FASEB J 1: 358–364
Halliwell B (1989 a) Superoxide, iron, vascular endothelium and reperfusion injury. Free Radic Res Commun 5: 315–318
Halliwell B (1989 b) Protection against tissue damage in vivo by desferrioxamine. What is its mechanism of action? Free Radic Biol Med 7: 645–651
Halliwell B (1990) How to characterize a biological antioxidant. Free Radic Res Commun 9: 1–32
Halliwell B, Aruoma OI (1991) DNA damage by oxygen-derived species. Its mechanism and measurement in mammalian systems. FEBS Lett 281: 9–19
Halliwell B, Grootveld M (1987) The measurement of free radical reactions in humans. FEBS Lett 213: 9–14
Halliwell B, Gutteridge JMC (1984) Lipid peroxidation, oxygen radicals, cell damage and antioxidant therapy. Lancet i: 1396–1398
Halliwell B, Gutteridge JMC (1986) Oxygen free radicals and iron in relation to biology and medicine: some problems and concepts. Arch Biochem Biophys 246: 501–504
Halliwell B, Gutteridge JMC (1989) Free radicals in biology and medicine, second edition. Clarendon, Oxford.
Halliwell B, Gutteridge JMC (1990 a) Role of free radicals and catalytic metal ions in human disease: an overview. Methods Enzymol 186: 1–85
Halliwell B, Gutteridge JMC (1990 b) The antioxidants of human extracellular fluids. Arch Biochem Biophys 280: 1–8
Halliwell B, Aruoma OI, Mufti G, Bomford A (1988 a) Bleomycin-detectable iron in serum from leukaemic patients before and after chemotherapy. FEBS Lett 241: 202–204

Halliwell B, Aruoma OI, Wasil M, Gutteridge JMC (1988b) The resistance of transferrin, lactoferrin and caeruloplasmin to oxidative damage. Biochem J 256: 311–312

Halliwell B, Grootveld M, Kaur H, Fagerheim I (1988c) Aromatic hydroxylation and uric acid degradation as methods for detecting and measuring oxygen radicals in vivo and in vitro. In: Rice-Evans C, Halliwell B (eds) Free radicals: methodology and concepts. Richelieu, London, pp 33–59

Halliwell B, Chirico S, Kaur H, Aruoma OI, Grootveld M, Blake DR (1992) Application of new assays for measuring free radical production to human rheumatoid patients. In: Davies KJA (ed) Oxidative damage and repair. Proceedings of the Pasadena conference. Pergamon, New York pp 846–855

Hayaishi M, Slater TF (1986) Inhibitory effects of Ebselen on lipid peroxidation in rat liver microsomes. Free Radic Res Commun 2: 179–185

Heiskala H, Gutteridge JMC, Westermarck T, Alanen T, Santavuori P (1988) Bleomycin-detectable iron and phenanthroline-detectable copper in the cerebrospinal fluid of patients with neuronal ceroid-lipofuscinoses. Am J Med Genet (Suppl) 5: 193–202

Hirsch EC, Brandel JP, Galle P, Javoy-Agid F, Agid Y (1991) Iron and aluminium increase in the substantia nigra of patients with Parkinson's disease: an X-ray microanalysis. J Neurochem 56: 446–451

Hope BT, Michael GJ, Knigge LM, Vincent SR (1991) Neuronal NADPH diaphorase is a nitric oxide synthetase Proc Natl Acad Sci USA 88: 2811–2814

Imlay JA, Linn S (1988) DNA damage and oxygen radical toxicity. Science 240: 1302–1309

Kanner J, German JB, Kinsella JE (1987) Initiation of lipid peroxidation in biological systems. Crit Rev Food Sci Nutr 25: 317–374

Kehrer JP (1989) Concepts related to the study of reactive oxygen and cardiac reperfusion injury. Free Radic Res Commun 5: 305–314

Komulainen H, Bondy SC (1988) Increased free intracellular Ca^{2+} by toxic agents; an index of potential neurotoxicity? Trends Pharmacol Sci 9: 154–156

Kyle ME, Nakae D, Sakaida I, Miccadei S, Farber JL (1988) Endocytosis of superoxide dismutase is required in order for the enzyme to protect hepatocytes from the cytotoxicity of hydrogen peroxide. J Biol Chem 263: 3784–3789

Langston JW, Irwin I, Ricaurte GA (1987) Neurotoxins, Parkinsonism and Parkinson's disease. Pharmcol Ther 32: 19–49

Laughton MJ, Halliwell B, Evans PJ, Hoult JRS (1989) Antioxidant and pro-oxidant actions of the plant phenolics quercetin, gossypol and myricetin. Effects on lipid peroxidation, hydroxyl radical generation and bleomycin-dependent damage to DNA. Biochem Pharmacol 38: 2859–2865

Laurindo FRM, da Luz PL, Vint L, Rocha TF, Jaeger RG, Lopes EA (1991) Evidence for superoxide radical-dependent coronary vasospasm after angioplasty in intact dogs. Circulation 83: 1705–1715

Liccione JJ, Maines MD (1988) Selective vulnerability of glutathione metabolism and cellular defence mechanisms in rat striatum to manganese. J Pharmacol Exp Ther 247: 156–161

Lim KH, Connolly M, Rose D, Siegman F, Jacobowitz I, Acinepura A, Cunningham JN Jr (1986) Prevention of reperfusion injury of the ischemic spinal cord: use of recombinant superoxide dismutase. Ann Thorac Surg 42: 282–286

Link EM (1988) The mechanism of pH-dependent hydrogen peroxide cytotoxicity in vitro. Arch Biochem Biophys 265: 362–372

Maly FE (1990) The B-lymphocyte: a newly-recognized source of reactive oxygen species with immunoregulatory potential. Free Radic Res Commun 8: 143–148

Martinovits G, Melamed E, Cohen O, Rosenthal J, Uzzan A (1986) Systemic administration of antioxidants does not protect mice against the dopaminergic neurotoxicity of 1-methyl-4-phenyl-1,2,5,6-tetrahydropyridine (MPTP). Neurosci Lett 69: 192–197

McConkey DJ, Hartzell P, Nicotera P, Wyllie AH, Orrenius S (1988) Stimulation of endogenous endonuclease activity in hepatocytes exposed to oxidative stress. Toxicol Lett 42: 123–130

McCord JM, Russell WJ (1988) Superoxide inactivates creatine phosphokinase during reperfusion of ischemic heart. In: Cerutti PA, Fridovich I, McCord JM (eds) Oxy-radicals in molecular biology and pathology. Liss, New York, pp 27–35

Meier B, Radeke H, Selle S, Raspe HH, Sies H, Resch K, Habermehl GG (1990) Human fibroblasts release reactive oxygen species in response to treatment with synovial fluids from patients suffering from arthritis. Free Radic Res Commun 8: 149–160
Meldrum B (1985) Excitatory amino acids and anoxic/ischaemic brain damage. Trends Neurosci 8: 47–48
Meneghini R (1988) Genotoxicity of active oxygen species in mammalian cells. Mutat Res 195: 215–230
Michel HS, Vaishnav YN, Kempski O, von Lubitz D, Weiss JF, Feuerstein G (1987) Breathing 100% oxygen after global brain ischemia in Mongolian gerbils results in increased lipid peroxidation and increased mortality. Stroke 18: 426–430
Muakkassah-Kelly SF, Andresen JW, Shih JC, Hochstein P (1982) Decreased (^{3}H)-serotonin and (^{3}H)-assat-spiperone binding consequent to lipid peroxidation in rat cortical membranes. Biochem Biophys Res Commun 104: 1003–1010
Muller A, Cadenas R, Graf P, Sies H (1984) A novel biologically active seleno-organic compound: I. Glutathione peroxidase activity in vitro and anti-oxidant capacity of PZ51 (Ebselen). Biochem Pharmacol 33: 3235–3240
Murrell GAC, Francis MJO, Bromley L (1990) Modulation of fibroblast proliferation by oxygen free radicals. Biochem J 265: 659–665
Pall HS, Williams AC, Blake DR, Lunec J, Gutteridge JMC, Hall M, Taylor A (1987) Raised cerebrospinal-fluid copper concentrations in Parkinson's disease. Lancet ii: 238–241
Patt A, Harkam AH, Burton LK, Rodell TC, Piermattei D, Schorr WJ, Parker NB, Berger EM, Horesh IR, Terada LS, Linas SL, Cheronic JC, Repine JE (1988) Xanthine oxidase-derived hydrogen peroxide contributes to ischemia reperfusion-induced edema in gerbil brains. J Clin Invest 81: 1556–1562
Pellegrini-Giampietro DE, Cherichi G, Alesiani M, Carla V, Moroni F (1988) Excitatory amino acid release from rat hippocampal slices as a consequence of free-radical formation. J Neurochem 51: 1961–1963
Puppo A, Halliwell B (1988) Formation of hydroxyl radicals from hydrogen peroxide in the presence of iron. Is haemoglobin a biological Fenton catalyst? Biochem J 249: 185–190
Puppo A, Halliwell B (1989) Oxidation of dimethylsulphoxide to formaldehyde by oxyhaemoglobin and oxyleghaemoglobin in the presence of hydrogen peroxide is not mediated by "free" hydroxyl radicals. Free Radic Res Commun 5: 277–281
Quinlan GJ, Halliwell B, Moorhouse CP, Gutteridge JMC (1988) Action of lead(II) and aluminium(III) ions on iron-stimulated lipid peroxidation in liposomes, erythrocytes and rat liver microsomal fractions. Biochim Biophys Acta 962: 196–200
Radi R, Beckman JS, Bush KM, Freeman BA (1991) Peroxynitrite oxidation of sulfhydryls. The cytotoxic potential of superoxide and nitric oxide. J Biol Chem 266: 4244–4250
Riederer P, Sofic E, Rausche WD, Schmidt B, Reynolds GP, Jallinger K, Youdim MBH (1989) Transition metals, ferritin, glutathione and ascorbic acid in Parkinsonian brains. J Neurochem 52: 515–520
Rosen AD, Frumin NV (1979) Focal epileptogenesis following intracortical haemoglobin injection. Exp Neurol 66: 277–284
Sarafian T, Verity MA (1991) Oxidative mechanisms underlying methyl mercury neurotoxicity Int J Dev Neurosci 9: 147–153
Saran M, Michel C, Bors W (1990) Reaction of NO with O_2^-. Implications for the action of endothelium-derived relaxing factor. Free Radic Res Commun 10: 221–226
Schor NF (1988) Inactivation of mammalian brain glutamine synthetase by oxygen radicals. Brain Res 456: 17–21
Shapira AHV, Mann VM, Cooper JM, Dexter D, Daniel SE, Jenner P, Clark JB, Marsden CD (1990) Anatomic and disease specificity of NADH CoQ_1 reductase (complex I) deficiency in Parkinson's disease. J Neurochem 55: 2142–2145
Sies H (1991) Oxidative stress, oxidants and antioxidants. Academic Press, New York
Siesjö BK (1988) Mechanisms of ischemic brain damage. Crit Care Med 16: 954–963
Siesjö BK (1990) Calcium, excitotoxins and brain damage. NIPS 5: 120–125
Siesjö BK, Bendek G, Koide T, Westerberg E, Wieloch T (1985) Influence of acidosis on lipid peroxidation in brain tissues in vitro. J Cereb Blood Flow Metab 5: 153–158

Singer TP, Trevor AJ, Castagnoli N Jr (1987) Biochemistry of the neurotoxic action of MPTP. Trends Biochem Sci 12: 266–270

Sofic E, Paulus W, Jellinger K, Riederer P, Youdim MBH (1991) Selective increase of iron in substantia nigra zona compacta of Parkinsonian brains. J Neurochem 56: 978–982

Sotomatsu A, Nakano M, Hirai S (1990) Phospholipid peroxidation induced by the catechol-Fe^{3+} (Cu^{2+}) complex: a possible mechanism of nigrostriatal cell damage. Arch Biochem Biophys 283: 334–341

Spector R, Eells J (1984) Deoxynucleoside and vitamin transport into the central nervous system. Fed Proc 43: 196–200

Stankovic A, Mitrovic DR (1991) Aluminium salts stimulate luminol-enhanced chemiluminescence production by human neutrophils. Free Radic Res Commun 14: 47–55

Steele JA, Stockbridge N, Maljkovic G, Weir B (1991) Free radicals mediate actions of oxyhemoglobin on cerebrovascular smooth muscle cells. Circulation Research 68: 416–423

Stocks J, Gutteridge JMC, Sharp RJ, Dormandy TL (1974) The inhibition of lipid autoxidation by human serum and its relationship to serum proteins and alpha-tocopherol. Clin Sci 47: 223–233

Touati D (1989) The molecular genetics of superoxide dismutase in *E. coli*. An approach to understanding the biological role and regulation of SODs in relation to other elements of the defence system against oxygen toxicity. Free Radic Res Commun 8: 1–9

Turrens JF, Crapo JD, Freeman BA (1984) Protection against oxygen toxicity by intravenous injection of liposome-entrapped catalase and superoxide dismutase. J Clin Invest 73: 87–95

Turski L, Bressler K, Rettig KJ, Loschman PA, Wachtel H (1991) Protection of substantia nigra from MPP^+ neurotoxicity by N-methyl-D-aspartate antagonists. Nature 349: 414–418

Van Loon APGM, Pesold-Hurt B, Schatz G (1986) A yeast mutant lacking mitochondrial manganese-superoxide dismutase is hypersensitive to oxygen. Proc Natl Acad Sci USA 83: 3820–3824

Watson BD, Ginsberg MD (1988) Mechanism of lipid peroxidation potentiated by ischemia in the brain. In Oxygen radicals and tissue injury. (Halliwell B ed) FASEB, pp 81–91

Weiner WJ, Nausieda PA, Klawans HL (1977) Effect of chlorpromazine on central nervous system concentrations of manganese, iron and copper. Life Sci 20: 1181–1186

Willmore LJ, Ballinger WE, Boggs W, Sypert GW, Rubin JJ (1980) Dendritic alterations in rat isocortex within an iron-induced chronic epileptic focus. Neurosurgery 7: 142–146

Willmore LJ, Triggs WJ, Gray JD (1986) The role of iron-induced hippocampal peroxidation in acute epileptogenesis. Brain Res 382: 422–426

Youdim MBH (1988 a) Brain iron. Neurochemical and behavioural aspects. Taylor and Francis, New York

Youdim MBH (1988 b) Iron in the brain: implications for Parkinson's and Alzheimer's diseases. Mt Sinai J Med 55: 97–101

Protection Against Peroxidation by Radicals in Cerebral Capillaries and Microvessels During Aging

J. M. Bourre

Summary

Nerve tissue (like cerebral capillaries and microvessels) has a high concentration of polyunsaturated fatty acids of the linoleic and alpha-linolenic series. It also consumes large amounts of oxygen. Peroxidation of these fatty acids due to free radicals is a substantial risk for the biochemistry and physiology of these membranes and can result in cellular death.

Systemic injection in the rat of substances that generate free radicals seriously alters various tissues (as shown by the production of conjugated dienes and malondialdehyde) but has only a slight effect on the brain. This shows that the blood-brain barrier confers considerable protection. In fact cerebral capillaries and microvessels show superoxide dismutase, glutathione peroxidase, and catalase activities that are higher than those in the rest of cerebral tissues during aging. These activities vary little, except for catalase, which decreases. Curiously, the concentrations of Mn, Cu, and Zn are not related to the enzyme activities, whereas these trace elements are necessary for the activities of superoxide dismutase. On the other hand, during aging the cerebral capillaries and microvessels undergo pronounced changes at the level of the polyunsaturated fatty acids: for example, the concentration of arachidonic acid (expressed as a percentage) decreases by half.

Introduction

Oxygen is absolutely necessary for life, but its use leads indirectly to the production of free radicals and peroxides (Ferradini 1986), which are toxic for the cells and their environment (Harmann 1981, Tappel 1980, Roth 1979). Free radical attacks are certainly toxic for the organism. It has been estimated that the duration of human life could be extended by 5 – 10 years if the peroxides and molecules that generate free radicals were controlled in the environment and in food (Roth 1979). Attack by free radicals very likely perturbs all levels of cellular structure and metabolism, and there is certainly a very marked effect on the membranes (Brizzee et al. 1984).

The central nervous system, principal consumer of oxygen in mammals, is an organ whose cell membranes are particularly rich in polyunsaturated fatty acids sensitive to oxidation reactions. Because neuronal renewal is nil, the integrity of nerve tissue depends closely on its ability to be protected against aggression by free radicals. The study of systems for protection against free radicals is today considered a very promising approach, both in neurotoxicology and in the search for new therapeutic strategies for the prevention and treatment of cerebral aging, including aging of vascular origin (post-ischaemic infarction) and perhaps also in Alzheimer's disease.

The presence of a generator of free radicals in the brain leads to cellular destruction, oedema, followed by death of the surviving cells and increase in vascular permeability. An important mechanism implicated in cellular death is lipid peroxidation, the consequence of attack by free radicals. The initial event is lysis of membrane phospholipids. This can occur directly, by radical attack, or indirectly, by activation of phospholipases (Cahn et al. 1983).

Phospholipase A_2 controls the deacylation of polyunsaturated fatty acids at position 2 on phospholipids, and this enzyme is activated by free radicals. These thus lead to the appearance of arachidonic acid, which is directly toxic, and is itself peroxidized to derivatives that are very toxic. Moreover, the metabolic derivatives of arachidonic acid synthesized by the cyclooxygenase pathways (the prostaglandins) and the lipoxygenase pathways (the leukotrienes) are indirectly harmful. They provoke vasoconstriction and ischaemia, or they accelerate the inflammatory process and, in consequence, generation of free radicals, thus perpetuating the vicious circle.

Peroxidation of membrane lipids perturbs enzyme activities, receptor-ligand relations, transport, and the asymmetry of the lipid bilayer. The links between proteins and lipids formed by the residues of peroxidation of fatty acids give rise to macromolecules that are not broken down in the lysosomes and accumulate to form lipofuscin.

Whatever the mechanism implicated, attacks on the cell can, without killing it, lead to the destruction of certain of its organelles. The destruction of the mitochondria releases in the cell elements of the respiratory chain: reduction of oxygen is then only partial, resulting in the appearance of free radicals. The mitochondria are directly implicated in the phenomenon of aging and of cellular death, through the generation of free radicals (Harmann 1981). Moreover, enzymes such as monoamine oxidase, xanthine oxidase, glycine oxidase, and others release hydrogen peroxide which in the presence of metals leads to the production of free radicals. Thus, though they can be formed in numerous ways, the main sources of superoxides are the electron transport chains of the mitochondria and microsomes.

Position 2 of phospholipids is usually occupied by a polyunsaturated fatty acid, most often 20:4 n-6, 22:4 n-6, 22:5 n-3, or 22:6 n-3. Works dealing with the influence of n-6 polyunsaturated fatty acids on the structure and function of the nervous system are very numerous. But the n-3 series polyunsaturated fatty acids play a very particular role in the membranes, especially in the nervous system; all the cerebral cells and organelles are extremely rich in this fatty acid (Bourre 1989).

Cerebral polyunsaturated fatty acids are products of elongation and desaturation of precursors that are obligatorily dietary (linoleic and alpha-linolenic acids). There is a relationship between the nature of dietary lipids and the accumulation of fatty pigments in the hippocampus. However, the brain controls the flow of dietary polyunsaturated fatty acids better than any other organ.

One of the causes of aging could be a loss of membrane asymmetry due to radical attack (Dobretsov et al. 1977; Shroeder 1984). Out of 100 membrane receptors studied in several tissues during aging, including human tissue, 75 % showed reduced affinity, 25 % were unchanged and 5 % showed an increase. It is estimated that in the course of life, there is a loss of from 50 % to 70 % of receptors and 25 % of enzyme activities (Roth 1979).

In addition, glutathione peroxidase is an exceptional enzyme that contains, selenium and has special kinetics (Chaudière and Tappel 1983). The decomposition of H_2O_2 by catalase is limited by the compartmentalization of this enzyme in the peroxisomes and by its weak affinity for its substrate. Thus, in fact, glutathione peroxidase degrades the major part of H_2O_2 in the cytosol and in the mitochondria. The activity of this enzyme decreases during aging, as does that of glutathione reductase (Danh et al. 1983; Vanella et al. 1982). In certain species, longevity is directly proportional to the activity of enzymes that protect against free radicals.

The main objective or our research programme is to identify the various components of the protective system of the brain, the cerebellum, and the cerebral capillaries. These components are: enzymes (selenium glutathione peroxidase, glutathione reductase, and catalase), intramembrane protectors (vitamin E), and reducing agents (glutathione and ascorbate).

If the production of free radicals is measured by electron paramagnetic resonance, it is found that they increase with age in all organs examined (Fig. 1), but in a more marked manner in the brain than in heart or liver, for example (Sawada and Carlson1987). In fact, nerve tissue has the second highest concentration of lipids, just after adipose tissue, but its lipids are structural, membrane lipids, formed of large amounts of polyethylene fatty acids and very sensitive to oxidation. Moreover, oxygen consumption by the brain is very high.

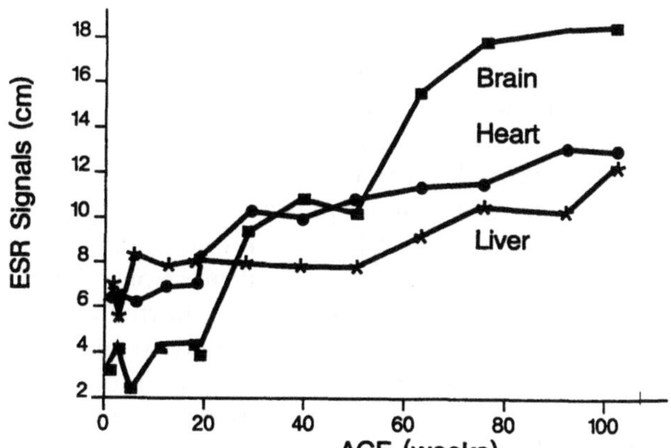

Fig. 1. Superoxide radicals produced during aging, measured by electron paramagnetic resonance. **ESR,** electron spin resonance

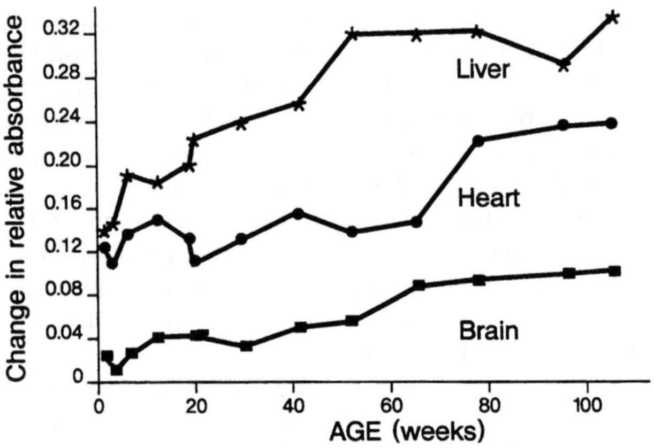

Fig. 2. Peroxidation index of lipids during aging in various tissues, quantified by measurement of conjugated dienes

On the other hand, if the results of peroxidation are measured by conjugated dienes (Fig. 2), it appears that peroxidation is lower in the brain than in other organs (Sawada and Carlson 1987). Consequenctly, the free radicals that are produced are rapidly neutralized, indicating a very high protection of nerve structures.

Results

None of the systemic intoxication models tested to date (cysteine, styrene, ferrocene, MEKP, paraquat, ethylmorphine, or phenylethylamine) have provided a very high neurotoxicity/hepatotoxicity ratio because of the effectiveness of the blood-brain barrier. The cerebral systems of protection against peroxidation are thus extraordinarily resistant and remain to be determined.

The necessity for extensive protection against oxidative alteration is clearly suggested by the fatty acid composition of cerebral capillaries. Some 32 % of total lipids are polyunsaturated fatty acids. Arachidonic acid and cervonic acid are the main polyunsaturated fatty acids present. The concentration of arachidonic acid is higher than that observed in the brain and cerebellum. Levels of linoleic acid, very low in brain, are about fivefold higher in cerebral capillaries (Tayarani et al. 1987).

The chromatographic elution volume using high-pressure liquid chromatography of a homogenate of cerebral capillaries shows a single peak of glutathione peroxidase activity that is eluted with a partition coefficient very close to that of liver selenium glutathione peroxidase. Their molecular weights are thus very similar. The glutathione peroxidase activity is exclusively due to selenium glutathione peroxidase (100 % inhibition by mercaptosuccinate).

The potential for enzymatic protection of cerebral capillaries against free radicals and hydroperoxide is high. Enzyme activities are greater than in brain, about two fold for glutathione peroxidase and glutathione reductase, threefold for catalase and superoxide dismutase (Table 1).

In cerebral capillaries during aging, levels of glutathione peroxidase, glutathione reductase, and superoxide dismutase show practically no change, in

Table 1. Comparison of specific activities of enzymes that protect against peroxidation in homogenates of rat cerebrum, cerebellum, and brain microvessels

Tissue	Glutathione peroxidase	Glutathione reductase	Catalase	Superoxide dismutase
Microvessels	47.6	48.4	4.8	48.0
Cerebrum	24.0	25.0	1.3	58.6
Cerebellum	24.0	23.7	1.6	57.7

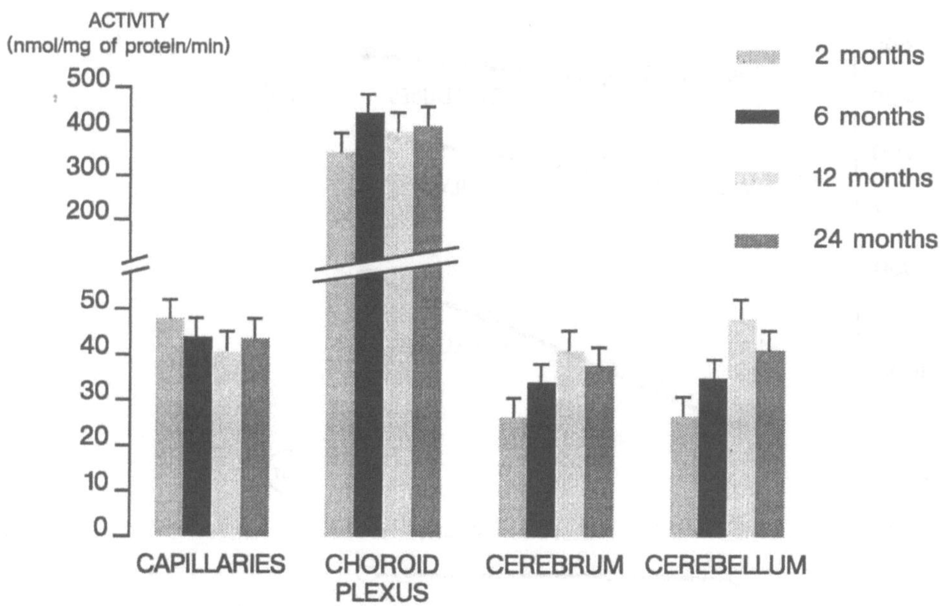

Fig. 3. Activities of various enzymes that protect against peroxidation during aging in different nerve structures

contrast to levels of catalase, which decrease (Tayarani et al. 1989) (Fig. 3). On the other hand, in brain, glutathione peroxidase and glutathione reductase increase up to 12 months and decrease slightly thereafter, while catalase remains stable and superoxide dismutase decreases by 100 % between 2 months and 2 years. It is interesting to note that the choroid plexus has specific activities similar to those of cerebral capillaries and shows the same changes during aging, except for glutathione peroxidase, whose levels are 10-fold higher. The significance of this is not known.

These results suggest that stimulation of the activity of glutathione peroxidase and induction of glutathione synthesis could very significantly reinforce the protective capacities of cerebral capillaries and thus of the brain.

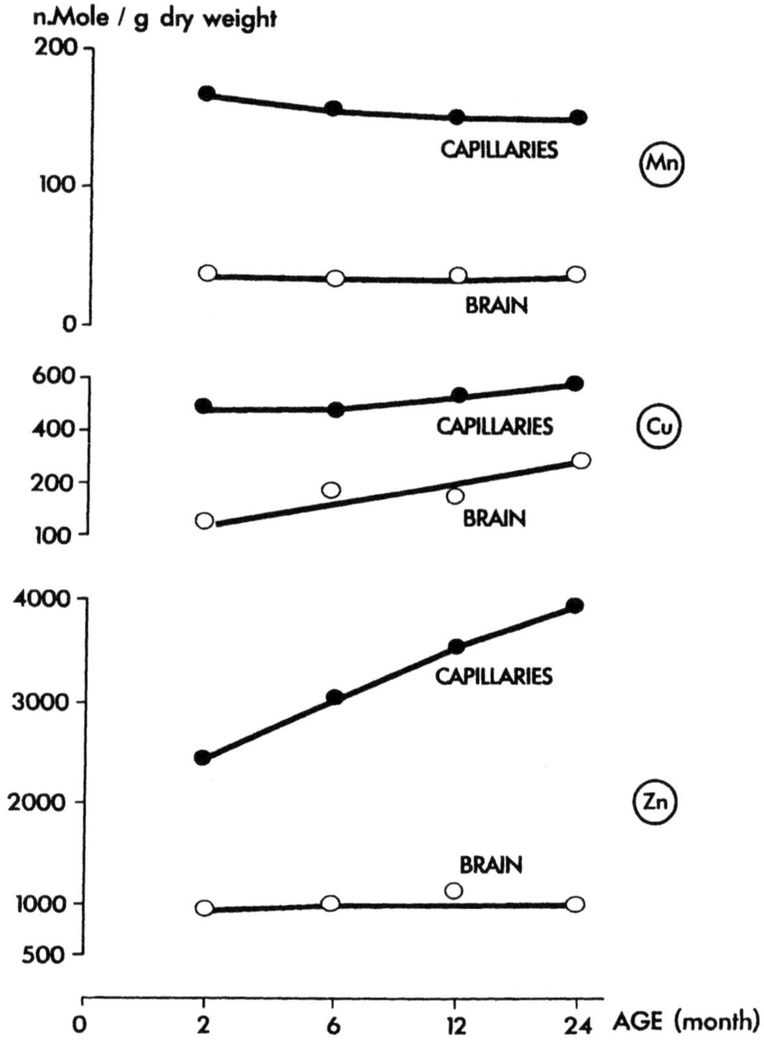

Fig. 4. Variation in trace element levels in cerebral capillaries and microvessels during aging. Materials and methods as described previously (Tayarani et al. 1989)

Superoxide dismutase requires the trace elements Mn, Cu, and Zn. However, variation in the levels of these in cerebral capillaries during aging does not parallel enzyme activities. In fact, cerebral capillaries contain much more Mn, Cu, and Zn that the brain (six-, four-, and fivefold more, respectively). During aging (Fig. 4), the Mn concentration does not vary either in the capillaries or in the brain. On the other hand, the level of Cu doubles in the brain whereas it remains stable in the capillaries, in contrast to Zn which increases in the capillaries but remains stable in the brain (Tayarani et al. 1989; Chaudière et al. 1988).

Conclusions

In the brain, copper and zinc superoxide dismutase (cytosolic), which represent 92% of total activity, decrease during aging, while manganese superoxide dismutase (mitochondrial) increases slightly. The variations in activity of superoxide dismutase during aging are interesting because these enzymes seem related to longevity in some species. It should be noted that protection systems are probably duplicated: a hereditary (genetic) deficiency of glutathione peroxidase hardly perturbs the organism, and a decrease in vitamin E does not lead to a compensatory increase in metabolism of the other protective systems.

Dietary polyunsaturated fatty acids control the structure and function of the brain (Bourre et al 1989) including those of capillaries and microvessels (Homayoun et al. 1988). However, at least in animals, during aging the uptake of arachidonic acid by the brain becomes more and more difficult (Gatti et al. 1986). It is more and more easily peroxidized (Mizuno et Ohta 1986), because among other reasons the enzymatic defenses against peroxidation decrease with age. Alteration of membrane polyunsaturated fatty acids leads to membrane dysfunction, which can be followed by cellular death.

The problem, then, is to avoid alterations in nerve tissue due to peroxidation during aging by trying to induce protective enzymes (Chaudièer et al. 1987), by increasing antioxidants in the membranes, and by facilitating their renewal, for example, that of vitamin E, whose presence in nerve structure plays a special role (Bourre et al. 1987; Clément et Bourre 1990; Vatassery et al. 1984; Machlin and Gabriel 1982). The increase in defenses at the level of the blood-brain barrier, in particular in the endothelial cells and cerebral capillaries and microvessels, is also an important objective.

References

Bourre JM (1989) Nature, origine et rôle des acides gras du système nerveux: un acide gras essentiel, l'acide alpha-linolénique, modula la structure et la fonction cérébrale. Bull. Acad. Natl. Med. 173: 1137–1151

Bourre JM, Clément M, Chaudière J (1987) Alteration of the alpha-tocopherol content in the brain and the peripheral nervous tissue of dysmyelinating mutants (Trembler, Quaking and Shiverer). Neurochem. Pathol 7: 91–97

Bourre JM, François M, Youyou A, Dumont O, Piciotti M, Pascal G, Durand G (1989) The effect of dietary alpha-linolenic acid on the composition of nerve membranes, enzymatic activity, amplitude of electrophysiological parameters, resistance to poisons and performance of learning tasks in rats. J. Nutr 119: 1880–1892

Brizzee KR, Eddy DE, Harman D, Mark Ordy J (1984) Free radical theory of aging: effect of dietary lipids on lipofuscin accumulation in the hippocampus of rats. Age 7: 9–15

Chan PH, Fishman RA, Caronna J, Schmidley JW, Prioleau G, Lee J (1983) Induction of brain edema following intracerebral injection of arachidonic acid. Ann. Neurol. (1989) 13: 625–632

Chaudière J, Tappel AL (1983) Purification and characterization of selenium glutathion peroxidase from hamster liver. Arch. Biochem. Biophys 226: 448–457

Chaudière J, Clément M, Gérard D, Bourre JM (1987) Induction of selenium glutathione peroxidase by stimulation of metabolic hydrogen peroxide production in vivo. Biochem Bioener 18: 247–256

Chaudière J, Clément M, Gérard D, Bourre JM (1988) Liver and brain alterations induced by vitamin E deficiency and intoxication with methyl ethyl ketone peroxide. Neurotoxicology 9: 173–180

Clément M, Bourre JM (1990) Alteration of α-tocopherol content in the developing an aging peripheral nervous system: persistence of high correlations with total and specific (n-6) polyunsaturated fatty acids. J. Neurochem 54: 21210–2117

Danh HC, Benedetti MS, Dostert P (1983) Differential changes in superoxide dismutase activity in brain and liver of old rats and mice. J. Neurochem 40: 1003–1007

Dobretsov GE, Borschezvskaya TA, Petrov VA, Vladimorov YA (1977) The increase of phospholipid bilayer rigidity after lipid peroxidation. FEBS Lett 84: 125–129

Ferradini C (1986) Espèces activées radicalaires de l'oxygène. Biochimie 68: 779–785

Gatti C, Noremberg K, Bruneti M, Tolato S, Calderini G, Gaiti A (1986) Turnover of palmitic and arachidonic acid in the phospholipids from different brain areas of adult and aged rats. Neurochem. Res 11: 241–252

Harman D (1981) The aging process. Proc Natl Acad Sci USA 78: 7124–7128

Homayoun P, Durand G, Pascal G, Bourre JM (1988) Alteration in fatty acid composition of adult rat brain capillaries and choroid plexus induced by a diet deficient in (n-3) fatty acids. Slow recovery by substitution with a non deficient diet. J. Neurochem 51: 45–48

Machlin LJ, Gabriel E (1982) Kinetics of tissue α-tocopherol uptake and depletion following administration of high levels of vitamine E. Ann. N.Y. Acad. Sci 393: 48–59

Mizuno Y, Ohta K (1986) Regional distributions of thiobarbituric acid reactive products, activities of enzymes regulating the metabolism of oxygen free radicals and some of the related enzymes in adult and age rat brains. J. Neurochem 46: 1344–1352

Roth GS (1979) Hormone receptor changes during adulthood senescence: significance for aging research. Fed. Proc.: 1910–1914

Sawada M, Carlson J (1987) Changes in superoxide radical and lipid peroxide formation in the brain heart and liver during the lifetime of the rat. Mech Ageing Dev 41: 125–137

Shroeder F (1984) Role of membrane lipid asymmetry in aging. Neurobiol. Aging 5: 448–457

Tappel AL (1980) Measurement and protection from in vivo lipid peroxidation. In: Pryor WA (ed) Free radicals in biology, Vol 4. Academic, New York, pp 2–47

Tayarani I, Chaudière J, Lefauconnier JM, Bourre JM (1987) Enzymatic protection against peroxidation damage in isolated brain capillaries. J. Neurochem 48: 1339–1402

Tayarani I, Cloez I, Clément M, Bourre JM (1989) Antioxydant enzymes and related trace elements in aging brain capillaries and choroid plexus. J. Neurochem 53: 817–824

Vanella A, Geremia E, d'Urso G, Tiriolo P, Di Silvestro I, Grimaldi R, Pinturo R (1982) Superoxide dismutase activities in aging rat brain. Gerontology 28: 108–113

Vatassery GT, Angerhofer CK, Knox CA, Deshmukh DS (1984) Concentration of vitamin E. in various neuroanatomical regions and subcellular fractions and the uptake of vitamin E by specific areas, of rat brain. Biochim. Biophys. Acta 792: 118–122

Antioxidant Protection of the Brain Against Oxidative Stress

V. E. Kagan, R. A. Bakalova, G. M. Koynova, V. A. Tyurin, E. A. Serbinova, V. V. Petkov, V. D. Petkov, D. S. Staneva, and *L. Packer*

Summary

Neural membranes are highly susceptible to oxidative damage and uncontrolled free radical oxidation is suggested to be involved in many neural and mental disorders. Vitamin E is known to be the major chain-breaking lipid-soluble antioxidant in the blood. This statement also holds true for neural membranes. Under physiological conditions low concentrations of vitamin E are sufficient to prevent membrane oxidative damage due to a unique ability of vitamin E molecules (tocopherols and tocotrienols) to act as membrane free radical harvesting centers by the mechanism of vitamin E regeneration (reduction of vitamin E radicals) by other intracellular reductants (NADPH- and NADH-dependent electron transport, vitamin C, reduced thiols). Vitamin E can synergistically interact with an unidentified protein kinase C-dependent mechanism, thus enhancing the overall antioxidant protection of brain membranes. Oxidative insult may result in a depletion of vitamin E in brain membranes followed by an accumulation of lipid peroxidation products. Enrichment of brain membranes with vitamin E by dietary supplementation provides for a higher protection of brain membranes against free radical oxidation. Thus the maintenance of a sufficient vitamin E concentration in the membranes of neural cells may be crucial for their resistance to oxidative stress. Ubiquinol, another lipid-soluble antioxidant, is less efficient than vitamin E (α-tocopherol) as an inhibitor of lipid peroxidation in brain membranes.

Introduction

There is an increasing body of experimental studies demonstrating the toxic effects of oxygen and organic free radicals in biological systems exposed to free radicals that have been formed endogenously or that result from external influences. Free radicals are known to be involved in different human disease processes. The evidence for free radical involvement includes:

1. abnormal free radical production in the disease,
2. deliberate application of free-radical generating systems into the cellular locus responsible for the disease reproduces its manifestations, and

3. free radical scavengers and other antioxidants control facets of the disease process.

All of the key functions of neural cells – creation and maintenance of the transmembrane potential, recognition and subsequent transduction of signal, production and regulatory function of secondary messengers, release and uptake of neurotransmitters – are extremely sensitive to excessive generation of oxygen and peroxyl radicals (Halliwell and Gutteridge 1989; Hasegawa 1990; Satya-Murti et al. 1986). Accumulation of the end-product of free radical peroxidation, lipofuscin granules, is a characteristic response of neural cells to different kinds of oxidative stress and aging. Brain trauma and ischemia injury (Floyd 1990; Hall 1989), Parkinson's disease, Alzheimer's disease. Down's syndrome, schizophrenia, epilepsy (Sofic et al. 1991; Subbarao et al. 1990; Abbott et al. 1990), complications caused by neuroleptics and antidepressants (Cadet and Lohr 1989) and ageing (Ansari et al. 1989) are only a few of many examples of neural and mental diseases in which free radicals may be involved and where antioxidant therapy is considered to be very promising.

Neural cells have three main systems for the protection and repair under conditions of oxidative stress:

1. direct antioxidant enzymes (superoxide dismutase, catalase, peroxidases);
2. proteases and phospholipases activated by oxidative modification of membranes; and
3. lipid- and water-soluble antioxidants.

Systems 1 and 2 constitute the two lines of enzymic antioxidant defense, whereas system 3 is considered to be mainly non-enzymic and dependent on nutritional factors.

Oxygen free radicals are continuously being produced intracellularly by oxidation-reduction reactions. The intracellular concentrations of superoxide radical (the first product of one-electron oxygen reduction) and of hydrogen peroxide (the product of two-electron oxygen reduction) are normally under the very strict enzymic control of superoxide dismutase (SOD) and hydroperoxidases, catalase and peroxidase. Hydroxyl radical, the three-electron product of oxygen reduction, is too highly reactive and short-lived. Hence it cannot be under specific enzymic control. The steady-state concentrations of hydroxyl radicals and its organic counterparts – alkoxyl and peroxyl radicals – are regulated by free radical scavengers, small antioxidant molecules able to interact with oxygen or organic radicals in the aqueous or membrane phase. Thus the role of free radical scavengers in membrane protection against oxidative damage may become crucial at this stage of free radical process.

Vitamin E and the ubiquinones Q9 and Q10 in their reduced form (ubiquinols) are the two major lipid-soluble antioxidants in brain membranes, while ascorbate and reduced thiols are the two major cytosolic antioxidants in the neural cells (Ramasarma 1985; Sokol et al. 1984). In this chapter, we discuss the possible interactions between these main lipid- and water-soluble antioxidants which may enhance the overall antioxidant protection of the cell.

Effects of Vitamin E and Ubiquinols on In Vitro Lipid Peroxidation in Rat Brain Organelles

Both vitamin E (tocopherols and tocotrienols) and ubiquinols act as free radical scavengers by interacting with peroxyl radicals of lipids in the membrane hydrophobic interior:

$$\text{In-OH} + \text{R-OO}' \xrightarrow{K_1} \text{In-O}' + \text{ROOH} \qquad (1)$$

where InOH and In-O' are the antioxidant and its radical, respectively, and R-OO' and ROOH are peroxyl radical and lipid hydroperoxide, respectively. The efficiency of radical scavenging is primarily determined by the reactivity of an antioxidant in reaction 1. It has been reported that, in solution, tocopherols are about an order of magnitude more reactive with peroxyl radicals than ubiquinols (Burkalova et al. 1979; Naumov and Khrapova 1983). The rate constants of tocopherols and ubiquinols in reaction 1 are 3×10^6 and 1×10^5 $M^{-1}s^{-1}$, respectively. This suggests that vitamin E can also function as a more potent antioxidant than ubiquinols Q9 and Q10 in membrane systems. To test this suggestion we compared the efficiency of (Fe(II)+ascorbate)-induced lipid peroxidation in rat brain synaptosomes and mitochondria by α-tocopherol and ubiquinols Q9 and Q10. The data in Table 1 show that the difference in the antioxidant activity of vitamin E and ubiquinols is retained in brain synaptosomes: α-tocopherol produces 50% inhibition of peroxidation at concentrations one order of magnitude lower than ubiquinol Q10. Taking into account that vitamin E concentrations in the brain are higher than those of ubiquinols (Ramasarma 1985), we conclude that vitamin E may be a more efficient antioxidant in the brain than ubiquinols.

Table 1. Inhibition of Fe(II)+ascorbate-induced lipid peroxidation in rat brain synaptosomes and mitochondria by α-tocopherol and ubiquinol Q10[a]

Antioxidant	Mitochondria[b]	Synaptosomes[b]
α-Tocopherol	3.0×10^{-7}	2.3×10^{-6}
Ubiquinol Q10	4.5×10^{-5}	1.1×10^{-5}

[a] Lipid peroxidation was induced for 5 min at 37 °C
[b] Constants for 50% inhibition **(M).**

Effects of Vitamin E on Endogenous Levels of Lipid Peroxidation

Age-Related Differences in Endogenous Lipid Peroxidation Levels in the Brain Structures of the Rat

The products of free radical reactions are considered important factors responsible for the progressive membrane damage in the course of ageing (Ansari et al. 1989). However, there is insufficient information about the content of endogenous products of free radical reactions in functionally and morphologically differentiated brain regions in young and old animals and of the effect of vitamin E on it.

In our experiments on young sexually mature (4 months) and old (22 months) male Wistar rats, the levels of endogenous lipid peroxidation products (hydroperoxides with conjugated dienes and fluorescent end-products of peroxidation) were measured in the frontal cerebral cortex, hippocampus, striatum, hypothalamus and cerebellum. The effect of dietary supplementation with vitamin E (an oral dose of 20 mg/kg α-tocopherol daily for 30 days) on the endogenous lipid peroxidation in these brain regions was also studied. The level of endogenous lipid peroxidation products was significantly higher in all brain structures studied for old rats as compared with young rats. The difference was remarkable (more than 100% – 200%) in hypothalamus and hippocampus and less pronounced, although significant, in the frontal cortex, striatum and cerebellum (Fig. 1). Thirty days of dietary supplementation with vitamin E had no significant effect on the endogenous level of lipid peroxidation products in the brain structures of young rats but produced a pronounced decrease in the content of hydroperoxides and fluorescent lipid peroxidation end-products in old rats. In fact, there was no significant difference in the content of endogenous lipid peroxidation products in brain structures of young and old rats kept on Vitamin E-supplemented diet.

Iron-Induced Activation of Endogenous Lipid Peroxidation in the Brain

Administration of oxidants or other promoters of peroxidation (e.g., transition metals) provokes the development of oxidative stress, which causes an exhaustion of vitamin E reserves in peripheral tissues after which accumulation of peroxidation products occurs. Under normal conditions, the brain is known to be a "slow turnover" (20–30 weeks) tissue in terms of both enrichment with vitamin E and depletion of vitamin E when the animals are kept on vitamin E-supplemented or vitamin E-deficient diets, respectively (Vatassery et al. 1984; Muller and Goss-Sampson 1989). Brain membranes are relatively resistant to inducers of free radical peroxidation in vivo (Subbarao and Richardson 1990), and under the same oxidative conditions the depletion of vitamin E and activation of lipid peroxidation in the membrane structures of the brain are hard to achieve (Vatassery et al. 1984; Muller and Goss-Sampson 1989; Subbarao and Richardson 1990; Packer et al. 1979).

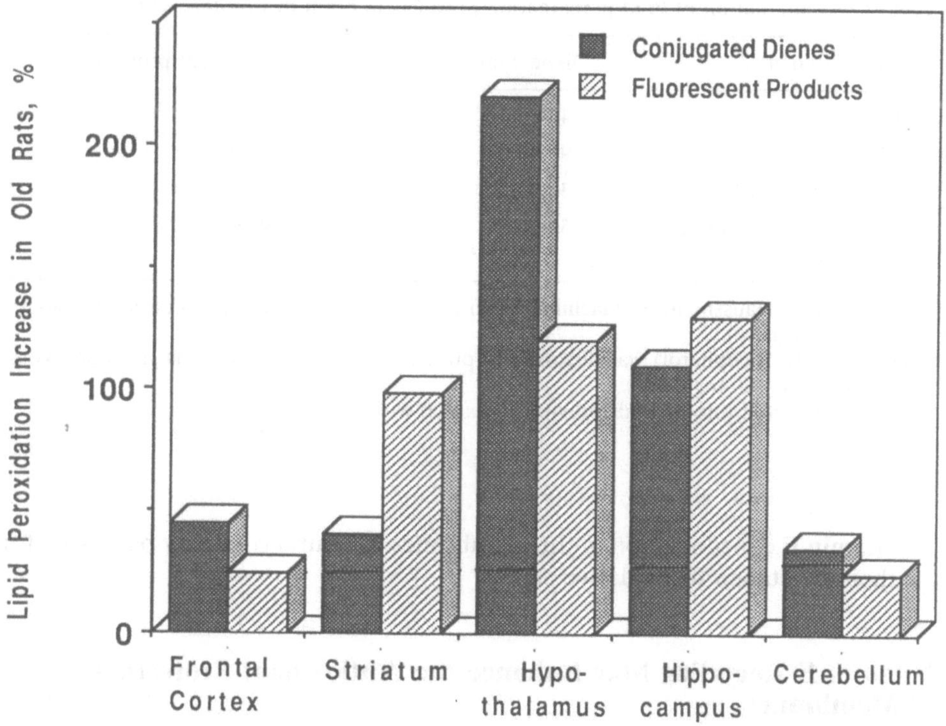

Fig. 1. Increase in the level of endogenous lipid peroxidation products in different brain structures of old (age 22 months) as compared to young (4 months) male Wistar rats

In our experiments we attempted to induce oxidative stress in the rat brain by repeated injections of iron, into the animals, which may act both as an initiator of free radical oxidation and as a catalyst of lipid hydroperoxide breakdown at the propagation stage. A single intramuscular injection of iron (in the form of Ferrum-Haussman, Pharmachim, 50 mg/kg body weight) into male Wistar rats (age 4 months) did not cause any accumulation of lipid peroxidation products in brain lipids, although significant activation of free radical oxidation was observed in the skeletal muscles and in the liver. However, after four injections of iron (once daily) in the same dose, a pronounced accumulation of endogenous lipid peroxidation was observed; the concentration of lipid hydroperoxides with conjugated double bonds was doubled and the level of fluorescent end-products of lipid peroxidation was 1.6 times higher than in the control (Table 2). This activation of endogenous lipid peroxidation was preceded by a Three-fold decrease in the concentration of vitamin E in the brain. Preliminary "saturation" of the animals with vitamin E (6 day duration, 50 mg daily) prevented iron-induced activation of lipid peroxidation.

The two experiments reported here indicate that

1. oxidative insult may result in vitamin E depletion followed by the accumulation of peroxidation products and
2. enrichment with vitamin E provides for a higher protection of the brain membranes against free radical oxidation. Thus the maintenance of sufficient

Table 2. Accumulation of lipid peroxidation products in brain of iron-loaded rats

Group of animals	Diene conjugates	Fluorescent products
Control	0.44±0.12	1.80±0.43
Iron-loaded[a]	0.97±0.09	2.90±0.30
Control + vitamin E[b]	0.38±0.19	0.94±0.16
Vitamin E-supplemented + iron-loaded[c]	0.50±0.14	1.30±0.26

[a] Iron (Ferrum-Haussman, Farmachim) was injected intramuscularly for 4 days (50 mg/kg body weight daily).
[b] Vitamin E (α-Tocopherol) was dissolved in purified sunflower oil and was given per os to rats for 6 days (50 mg daily).
[c] Iron was injected into animals previously loaded with vitamin E.

vitamin E concentration in the membranes of neural cells may be crucial for their resistance to oxidative stress.

Vitamin E Recycling May Enhance the Antioxidant Protection of Membranes

The steady-state concentrations of vitamin E in membranes are determined by (1) efficiency of its incorporation into membranes due to transfer from blood lipoproteins, and (2) its metabolism in membranes. The main intramembrane metabolic pathway of vitamin E is believed to be its scavenging lipid peroxyl and alkoxyl radicals in the course of initiation and propagation of lipid peroxidation (reaction 1). The radical of vitamin E, tocopheroxyl radical (Toc-O'), is formed in this reaction. Tocopheroxyl radical is much less reactive than other lipid radicals (peroxyl or alkoxyl radicals) generated in the course of lipid peroxidation. Thus vitamin E serves to break the chain of free radical reactions in lipid peroxidation. However, the tocopheroxyl radical is susceptible to oxidation or to destruction by reacting with itself or by other reactions that cause it to decompose as a result of radical initiated reactions:

$$\text{Toc-O'} + \text{Toc-O'} \rightarrow \text{Products} \quad (2)$$
$$\text{Toc-O'} + \text{L-OO'} \rightarrow \text{Products} \quad (3)$$

Unless reduced (regenerated) to its original antioxidant form by an appropriate reductant:

$$\text{Toc-O'} + \text{Red-H} \rightarrow \text{Toc-OH} + \text{Red'} \quad (4)$$

it will be lost. It is known from in vitro studies that ascorbate (vitamin C) can regenerate vitamin E from its free radical form (Packer et al. 1979). It has also been proposed that ubiquinols may serve to protect vitamin E against loss by interacting with its radical (Mellors and Teppel 1966). However, it is not known whether this is an important activity of membranes.

Electron Spin Resonance Assay of Tocopheroxyl Reactions and Recycling Activity in Natural Membranes

We have developed simple and convenient methods to generate phenoxyl radicals from vitamin E and its synthetic homologues based on their: (1) enzymatic oxidation by (lipoxygenase + polyunsaturated fatty acid, e.g., arachidonic, linolenic) system, or (2) non-enzymatic oxidation by azo-initiator of peroxyl radicals, 2,2'-azo-bis-(2,4-dimethylvaleronitrile) (AMVN) or (3) irradiation by UVB light which is directly absorbed by vitamin E molecules (Packer et al. 1989; Kagan et al. 1990a). Lipoxygenase generates peroxyl radicals of polyunsaturated fatty acids. AMVN is thermally decomposed to produce carbon-centered radicals which are rapidly converted into peroxyl radicals in the presence of oxygen. In these two systems peroxyl radicals interact with vitamin E to generate chromanoxyl radicals. Thus both systems imitate interaction of tocopherol with peroxyl radicals in the course of lipid peroxidation. The advantage of the UVB induction of vitamin E radicals is the absence of other radicals (peroxyl, alkoxyl) in the system which may interact with the recycling agents.

In the first experiments we used a homologue of α-tocopherol, 2,2,5,7,8-pentamethyl-6-hydroxy-chromane (α-C1), which is devoid of the phytol chain of vitamin E (Packer et al. 1989). This homologue is less hydrophobic than tocopherols and is more readily and uniformly distributed in the membranes. The steady-state concentrations of chromanoxyl radicals generated from α-C1 appeared to be high enough to use electron spin resonance (ESR) to follow reactions of chromanoxyl radicals in cell organelles. However, normal vitamin E concentrations in natural membranes (liver and brain microsomes, mitochondria) are too low to provide a detectable tocopheroxyl radical ESR signal. We also used other homologues of α-tocopherol differing in the length of their chain to enrich natural or liposomal membranes by their addition in vitro. We found that α-C6, an alpha-tocopherol homologue with 6-carbon side chain, gave well resolved ESR spectra of its chromanoxyl radicals.

Thus we had at our disposal simple and reliable methods of generating tocopheroxyl or homologous chromanoxyl radicals in measurable and quantifiable amounts in both natural membranes and in liposomes. With ESR we could simultaneously follow the time-course of changes in the amounts of these radicals. This allowed us to directly study membrane reactions of chromanoxyl radicals of vitamin E and of its homologues.

Recycling of Vitamin E and Its Homologues in Liposomes and Membranes

Using the methods described above we were able to demonstrate that ascorbate is efficient in regeneration of chromanoxyl radicals of tocopherol and its homologues not only in liposomes but also in human low density lipoproteins and rat tissue membrane preparations, i.e., brain and liver microsomes and mitochondria. Enzyme-dependent mechanisms which prevent accumulation of chromanoxyl radicals derived from vitamin E homologues were characterized in these membranes. NADPH or NADH in microsomes, as well as NADH or succinate in mitochondria, prevented accumulation of chromanoxyl radicals until these substrates were fully consumed (Fig. 2). Thus we concluded that rat liver and brain homogenates, microsomes and mitochondria have both enzymatic electron transport-dependent and non-enzymatic mechanisms for reducing chromanoxyl radicals.

We were able to demonstrate that ubiquinones synergistically enhance enzymic NADH- and NADPH-dependent recycling of tocopherols by electron transport in mitochondria and microsomes. In liposomes, reduced ubiquinone Q10, ubiquinol Q10, directly quenched ESR signal of tocopheroxyl radical, indicating that the reduction of tocopheroxyl radical by ubiquinol had occurred. Thus we conclude that antioxidant effects of ubiquinols (ubiquinones) are mainly due not to their direct radical scavenging reactivity, but rather to their ability to stimulate more efficient recycling of tocopherols interacting with electron-transport enzymes.

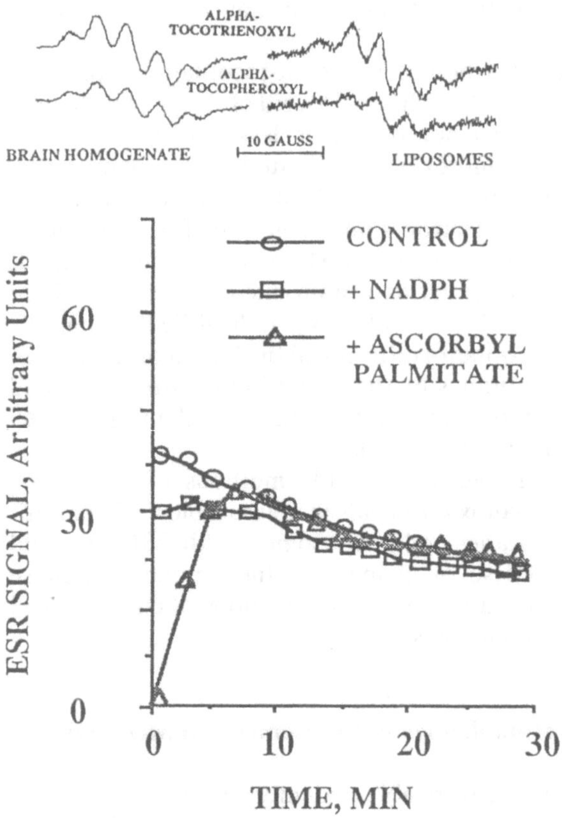

Fig. 2a, b. Typical electron spin resonance (ESR) spectra of vitamin E radical in rat brain homogenates and in liposomes from brain lipids. **b** Time course of α-tocopheroxyl ESR signal generated by lipoxygenase+linolenic acid in Sprague-Dawley rat brain homogenate and the effects of NADPH or ascorbyl palmitate

Under physiological conditions, low concentrations of vitamin E are sufficient to prevent membrane oxidative damage due to a unique ability of vitamin E molecules (tocopherols and tocotrienols) to act as membrane free radical harvesting centers. Location of the chromanol nucleus of vitamin E at the membrane interface affords efficient interaction with lipid radicals and regeneration of tocopheroxyl radicals. The antioxidant power of vitamin E is enhanced by other intracellular reductants (vitamin C, reduced thiols) or by NADPH- and NADH-dependent electron transport. Thus, the well-known synergistic effects of these physiologically important antioxidants (reductants) with vitamin E are mediated via their ability to donate electrons necessary for recycling tocopheroxyl radicals in membranes. Non-toxic reductants capable of enhancing the recycling efficiency of chromanoxyl radicals may lead to new avenues in clinical applications of vitamin E.

Protein Kinase C Inhibits Lipid Peroxidation in Brain Membranes and Synergistically Interacts with Vitamin E

Protein Kinase C-Dependent Inhibition of Lipid Peroxidation in Brain Membranes

The efficiency of many processes in the cell is controlled by secondary messengers generated by specific receptor-enzyme complexes. One of the general regulatory mechanisms is the protein kinase C-dependent pathway. Recently we showed that phorbol-12-myristate-13-acetate (PMA), the protein kinase C activator and a structural homologue of the secondary messenger diacylglycerol, produces a concentration-dependent lipid peroxidation inhibition in liver and muscle microsomes mediated by a protein kinase C-supported mechanism (Baldenkov et al. 1988). It was important to answer two questions:

1. whether this protein kinase C-mediated mechanism of lipid peroxidation regulation is operative in brain membranes, and
2. whether there is an interaction of protein kinase C-supported lipid peroxidation inhibition with vitamin E.

To answer these questions we studied the effects of PMA and of the protein kinase C inhibitor polymyxin B as well as of the adenylate cyclase activator forskolin on Fe(II)+ascorbate-dependent lipid peroxidation in brain cortex membrane fractions isolated from male Wistar rats (150–180 g). We found that PMA, preincubated with brain homogenates for 10 min at 30 °C prior to the isolation of membrane fractions, produced a concentration-dependent inhibition of the non-enzymic lipid peroxidation in plasma membranes, mitochondria, microsomes, and in the brain homogenate itself (Fig. 3). This inhibitory effect of PMA was significant at concentrations as low as 10^{-8} *M* and reached 50% inhibition of lipid peroxidation in the micromolar concentration range. Plasma membranes were least succeptible to PMA-dependent regulation of lipid peroxidation. In liposomes prepared from brain lipids, we did not observe any

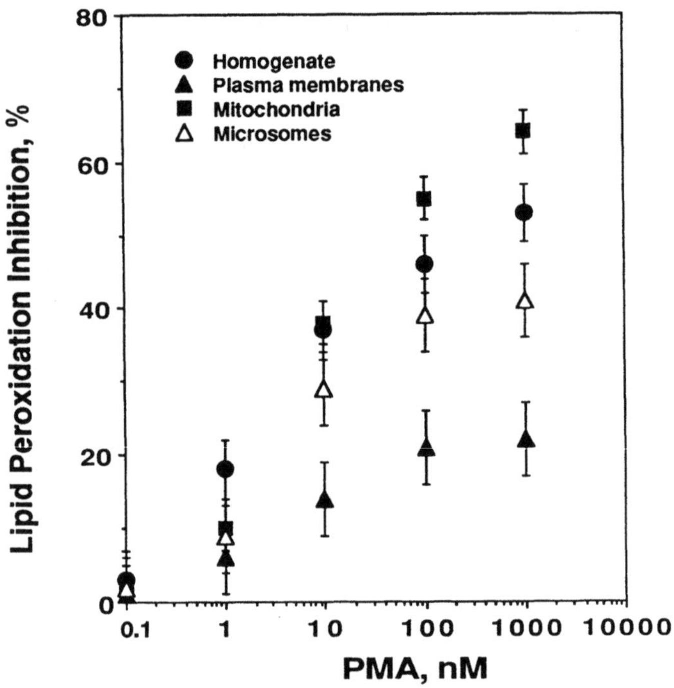

Fig. 3. Effect of phorbol-12-myristate-13-acetate (PMA) on Fe(II)+ascorbate-induced lipid peroxidation in homogenates, plasma membranes, mitochondria and microsomes isolated from male Wistar rat brain. PMA was added to the homogenate, incubated for 10 min at 30 °C, and membrane fractions were isolated. Lipid peroxidation was induced for 10 min at 37 °C and assayed by the TBA test

lipid peroxidation inhibition by PMA concentrations up to 10^{-5} *M*. This suggests that lipid peroxidation inhibition by PMA is not due to its direct radical scavenging effect but is rather mediated by some enzymic mechanism.

It is known that PMA at nanomolar concentrations exerts a stimulant effect on protein kinase C. This suggests that the effects of PMA on lipid peroxidation may be mediated by protein kinase C activation. The participation of protein kinase C in the inhibitory effect of PMA was substantiated by the experiments with PMA analogues, which are not efficient in protein kinase C activation: 4α-phorbol and 4α-phorbol-didecanoate. Neither compound inhibited lipid peroxidation in brain homogenates and membrane fractions (Table 3). The participation of protein kinase C in the PMA effect was further confirmed by the experiments with polymyxin B, a known inhibitor of protein kinase C. The inhibitory effect of PMA was substantially decreased in the presence of Polymyxin B. Polymyxin B itself did not exert a direct antioxidant effect in brain membranes or liposomes. In fact, a significant stimulation of lipid peroxidation by Polymyxin B was observed.

Protein kinase C- and cAMP-dependent protein kinases have opposite effects on several important physiological processes (e.g., cell proliferation, contrac-

Table 3. Effects of phorbol esters, polymyxin B and forskolin on Fe(II) + ascorbate-induced lipid peroxidation in rat brain membranes

Additions	Homogenate	Lipid peroxidation inhibition (%) Plasma Membranes	Mitochondria	Microsomes
PMA (0.1 μ*M*)[b]	50.0±4.0	45.2±4.1	27.0±3.3	32.7±3.0
Phorbol-di-decanoate (0.1 μ*M*)	0	0	0	0
Phorbol	0	2.7±0.1	0	2.2±0.1
Polymyxin B (0.1 m*M*)	–53.7±5.6[a]	–12.5±1.5[a]	–25.0±4.5	–23.7±5.3[a]
PMA (0.1 μM) + polymyxin B (0.1 m*M*)	18.5±3.2	19.0±2.5	6.3±0.7	– 9.7±1.7[a]
Forskolin (1 μ*M*)	0.5±0.1	17.7±4.1	2.5±0.4	10.5±3.3
Forskolin (1 μ*M*) + PMA (0.1 μ*M*)	34.5±3.3	39.5±7.5	12.6±2.4	19.0±2.2

[a] Stimulation of lipid peroxidation.
[b] PMA, phorbol-12-myristate-13-acetate.

tility, Ca2+channel activity, neutrophil activation). Does this fact hold true for the protein kinase C-dependent regulation of lipid peroxidation in the brain membranes? To answer this question we compared the effect of PMA with that of forskolin, an adenylate cyclase activator, on lipid peroxidation. The addition of forskolin and PMA to the membrane suspension and subsequent incubation for 10 min at 30 °C (which is sufficient for the adenylate cyclase activation) resulted in a significant decrease of the inhibitory effect of PMA (Table 3), although forskolin itself caused some inhibition of lipid peroxidation. We suggest that the stimulation of cAMP synthesis and the subsequent activation of cAMP-dependent protein kinases prevent the inhibitory effects of PMA on lipid peroxidation in brain membranes.

Although the specific mechanisms underlying the effects of PMA and forskolin on lipid peroxidation are still to be elucidated, these results suggest that lipid peroxidation in the brain membranes is regulated by the system of secondary messengers. PMA, an analogue of diacylglycerol, activating protein kinase C, inhibits lipid peroxidation, while forskolin, an activator of adenylate cyclase and cAMP-dependent reactions, prevents the inhibitory action of PMA.

Interaction of Protein Kinase C-Dependent Lipid Peroxidation Regulation with Vitamin E

One of the possibilities which may be considered to explain the PMA-dependent lipid peroxidation inhibition in brain membranes is its enhancement of the antioxidant effect of vitamin E by an increased efficiency of its enzymic regeneration. We report here results of experiments showing that PMA can synergistically interact with vitamin E in inhibiting lipid peroxidation in brain membranes. When both PMA (0.1 μ*M*) and α-tocopherol (10 μ*M*) were present in the suspension of rat brain microsomal membranes incubated in the presence of Fe(II)+ascorbate, no accumulation of lipid peroxidation products was found (100% inhibition of lipid peroxidation). PMA alone gave 40% inhibition of lipid peroxidation in microsomes at this concentration, while α-tocopherol suppressed lipid peroxidation by only 15%. Thus PMA and α-tocopherol can synergistically enhance their antioxidant potency in the brain membranes.

The interrelations between α-tocopherol and PMA suggest a regulatory process in which, at a normal or an increased content of vitamin E, protein kinase C is inhibited, i.e., down-regulated. In fact, down-regulation of protein kinase C by vitamin E has been recently reported (Mahoney and Azzi 1988; Boscoboinik et al. 1991). We speculate that the protein kinase C-dependent inhibition of lipid peroxidation is not turned on unless the concentration of vitamin E decreases due to its consumption in free radical reactions. Then the vitamin E-dependent inhibition of protein-kinase C disappears and its control over lipid peroxidation may become increasingly important in the overall regulation of the antioxidant defense (Kagan et al. 1990b).

Acknowledgements. Research reported in this chapter was supported by NIH grant CA 47597 and by the Bulgarian Academy of Sciences. The authors thank Eisai Co (Japan) for providing ubiquinone Q10.

References

Abbott LC, Nejad HH, Bottje WG, Hassan AS (1990) Glutathione levels in specific brain regions of genetically epileptic (tg/tg) mice. Brain Res Bull 25: 629–631

Ansari KA, Kaplan E, Shoeman D (1989) Age-related changes in lipid peroxidation and protective enzymes in the central nervous system. Growth Dev Aging 53: 117–121

Baldenkov GN, Serbinova EA, Bakalova RA, Tkachuk VA, Kagan VE, Stoytchev TS (1988) The role of secondary messengers in the regulation of lipid peroxidation in rat liver microsomes. Free Radic Res Commun 4: 277–281

Boscoboinik D, Szewczyk A, Hensey C, Azzi A (1991) Inhibition of cell proliferation by alpha-tocopherol. Role of protein kinase C. J Biol Chem 266: 6188–6194

Burlakova EB, Kukhtina EN, Olkhovskaya IP, Sarycheva IK, Sinkina EB, Khrapova NG (1979) Study of the antiradical activity of tocopherol analogues and homologues by the method of chemiluminescence. Biophysics 24: 989–993

Cadet JL, Lohr JB (1989) Possible involvement of free radicals in neuroleptic-induced movement disorders. Ann NY Acad Sci 570: 176–185

Floyd RA (1990) Role of oxygen free radicals in carcinogenesis and brain ischemia. FASEB J 4: 2587–2597

Hall ED (1989) Free radicals and CNS injury. Critl Care Clin 5: 793–805

Halliwell B, Gutteridge JMC (1989) Free radicals in biology and medicine, 2nd edn. Clarendon, Oxford

Hasegawa T (1990) Depressive effects of lipid peroxides mediated via a purine receptor. Biochem Pharmacol 40: 1463–1467

Kagan VE, Serbinova EA, Packer L (1990a) Generation and recycling of radicals from phenolic antioxidants. Arch Biochem Biophys 280: 33–39

Kagan V, Packer L, Serbinova E, Bakalova R, Stoyanovsky D, Zhelev Z, Harfouf, Kitanova S, Rangelova D (1990b) Mechanisms of vitamin E control of lipid peroxidation. In: Reddy CC,

Hamilton GA, Madastha KM (eds) Biological oxidation systems, vol 2. Academic, San Diego, pp 889–908
Mahoney CW, Azzi A (1988) Vitamin E inhibits protein kinase C activity. Biochem Biophys Res Commun 154: 694–697
Mellors A, Tappel AL (1966) The inhibition of mitochondrial peroxidation by ubiquinone and ubiquinol. J Biol Chem 241: 4353–435620
Muller DPR, Goss-Sampson MA (1989) Role of vitamin E in neural tissue. Ann NY Acad Sci 570: 146–155
Naumov VV, Khrapova NG (1983) Study of the interaction of ubiquinone and ubiquinol with peroxide radicals by the chemiluminescence method. Biophysics 28: 774–780
Packer JE, Slater TF, Willson RL (1979) Direct observation of a free radical interaction between vitamin E and vitamin C. Nature 278: 737–738
Packer L, Maguire JJ, Melhorn R, Serbinova E, Kagan V (1989) Mitochondria and microsomal membranes have a free radical reductase activity. Biochem Biophys Res Commun 159: 229–235
Ramasarma T (1985) Natural occurrence and distribution of coenzyme Q. In: Lenaz G (ed) Coenzyme Q. Wiley, New York, pp 67–81
Satya-Murti S, Howard L, Krohel G, Wolff B (1986) The spectrum of neurologic disorder from vitamin E deficiency. Neurology 36: 917–921
Sofic E, Paulus W, Jellinger K, Riederer P, Youdium MB (1991) Selective increase of iron in substantia nigra zona compacta of parkinsonian brains. J Neurochem 56: 978–982
Sokol RJ, Heubi JE, Innaccone ST, Bove KE, Balisteri WF (1984) Vitamin E deficiency with normal serum vitamin E concentrations in children with chronic cholestasis. N Engl J Med 310: 1209–1212
Subbarao KV, Richardson JS (1990) Iron-dependent peroxidation of rat brain: a regional study. J Neurosci Res 26: 224–23218
Subbarao KV, Richardson JS, Ang LC (1990) Autopsy samples of Alzheimer's cortex show increased peroxidation in vitro. J Neurochem 55: 342–345
Vatassery GT, Angerhofer CK, Knox CA, Desmukh DS (1984) Concentrations of vitamin E in various neuroanatomical regions and subcellular fractions, and the uptake of vitamin E by specific areas of rat brain. Biochim Biophys Acta 792: 118–122

Antioxidant Deficiency and Neurological Disease in Humans and Experimental Animals

D. P. R. Muller, M. A. Goss-Sampson, and *C. J. MacEvilly*

Summary

It is now recognised that vitamin E (α-tocopherol) is necessary for the maintenance of normal neurological structure and function, with a severe and chronic deficiency of vitamin E resulting in a characteristic progressive and crippling neurological disorder.

In recent years we have established a rat model of vitamin E deficiency in an attempt to answer fundamental questions such as, why should neural tissues be particularly at risk from a deficiency of vitamin E and what are the sequence of events and the mechanisms involved.

In summary, our studies in rats to date have shown, that during a deficiency of vitamin E:

1. Electrophysiological abnormalities are present which are very similar to those seen in vitamin E-deficient humans and which, therefore, validate the experimental model.
2. Neural tissues preferentially conserve vitamin E.
3. Other antioxidants do not compensate.
4. Indices of lipid peroxidation are increased.
5. Different neural regions show differing susceptibilities to in vitro free radical attack which are consistent with the characteristic neuropathology.
6. Axonal transport is decreased.

Introduction

Oxygen derived free radicals and/or impaired antioxidant defences have been implicated in an increasing number of disease processes, including neurological disorders (Halliwell and Gutteridge 1985, 1989). In vivo, vitamin E (α-tocopherol) is the only well-recognised, lipid-soluble chain-breaking antioxidant (Burton et al. 1983) and may, therefore, be expected to play an important role in protecting highly lipid structures such as the brain, spinal cord and peripheral nerves from free radical damage.

Since its discovery in 1922 by Evans and Bishop (1922), the role of vitamin E in human nutrition has been the subject of much dispute. In recent years it has become clear, however, that vitamin E is necessary for normal neurological structure and function. The evidence comes from four principal sources: from patients with 1. abetalipoproteinaemia, 2. other chronic disorders of fat absorption and 3. an isolated deficiency of vitamin E, and also 4. from comparative neuropathological studies in man and experimental animals.

This evidence will be considered in turn. Studies in a vitamin E-deficient rat model will then be described in which attempts have been made to answer fundamental questions such as, why neural tissues should be particularly at risk from vitamin E deficiency, what is the sequence of events taking place and what are the mechanisms involved.

Vitamin E Deficiency and Neurological Dysfunction in Man

Abetalipoproteinaemia

In this rare inherited disorder of lipoprotein metabolism serum concentrations of α-tocopherol are undetectable from birth (Kayden et al. 1965; Muller et al. 1974). This condition results from the dual effects of malabsorption and impaired transport. Chylomicrons (the main absorptive vehicle for α-tocopherol) cannot be formed, and there is a lack of betalipoprotein or low density lipoprotein (LDL), a major carrier protein. Abetalipoproteinaemia, therefore, provides a model for the study of the effects of vitamin E deficiency in humans.

Among the clinical features of abetalipoproteinaemia are an ataxic neuropathy and pigmentary retinopathy which typically develop during the second decade of life and have been described as "devastating" (Herbert et al. 1978). Since the vitamin E-deficient chick was known to develop a cerebellar disorder with ataxia (Pappenheimer and Goettsch 1931) and neurological lesions had been described in several other animal species (Wasserman and Taylor 1972), we decided to treat our patients with abetalipoproteinaemia with large oral doses of vitamin E (approximately 100 mg/kg/day of α-tocopheryl acetate; Ephynal, Hoffmann-La Roche and Company, Limited).

Eight patients who have been receiving such therapy have been followed for periods of 19 – 25 years. Treatment with vitamin E resulted in detectable serum concentrations of the vitamin, correction of the abnormal in vitro haemolysis associated with vitamin E deficiency, and either prevention (if started sufficiently early) or amelioration of the neurological and retinal disease (Muller et al. 1977; Muller and Lloyd 1982). Other investigators have also reported the beneficial effects of vitamin E therapy in this condition (Kane and Havel 1989; Azizi et al. 1978; Miller et al. 1980; Hegele and Angel 1985).

Other Disorders of Fat Malabsorption

Vitamin E deficiency can occur in any chronic disorder of fat absorption, but is likely to be most severe in conditions in which there is an interruption of the enterohepatic circulation of bile salts which are essential for the solubilisation and absorption of the vitamin (Harries and Muller 1971; Sokol et al. 1983). Neurological features similar to those found in abetalipoproteinaemia have now been described in patients with chloestatic liver disease (Rosenblum et al. 1981; Elias et al. 1981; Guggenheim et al. 1982; Alvarez et al. 1983; Sokol et al. 1985), extensive intestinal resection (Harding et al. 1982; Howard et al. 1982) and cystic fibrosis (Elias et al. 1981; Willison et al. 1985). Improvement in neurological function has been reported with appropriate treatment (Elias et al. 1981; Guggenheim et al. 1982; Sokol et al. 1985; Harding et al. 1982; Howard et al. 1982). In cholestatic liver disease it is generally necessary to give the vitamin by intramuscular injection (Harries and Muller 1971; Sokol et al. 1983, 1985; Guggenheim et al. 1982). Evidence for a causal relationship between vitamin E deficiency and the neurological sequelae was provided by the study of Sokol et al. (1985), who followed the effect of vitamin E supplementation in children with cholestatic liver disease. Two young patients (less than 3 years of age) remained normal after 18 months of therapy. Neurological function in three symptomatic children (also less than 3 years old) normalised after 18–32 months of treatment, and a group of older symptomatic children (5–17½ years) showed improvement (using a clinical scoring system) after 18–48 months of treatment.

Isolated Deficiency of Vitamin E

A number of patients have now been reported with a familial isolated deficiency of vitamin E without generalised fat absorption, who have similar neurological features to the patients described above with various fat malabsorptive disorders and have responded in a similar way to treatment (Sokol et al. 1988). Studies using deuterated tocopherols in these patients suggest that absorption is normal but that the disappearance from plasma is more rapid than from controls (Traber et al. 1990). It has been suggested that the patients lack a functional hepatic binding protein for α-tocopherol which is necessary for its transfer to very low density lipoproteins (VLDL). As a result the secretion of vitamin E back into the circulation is impaired and it is sequestered in the liver. There is no evidence for a deficiency of any other nutrient in these patients, and thus they provide further evidence for a causal relationship between a deficiency of vitamin E and the neurological sequelae.

Neuropathological Studies

The neuropathology associated with vitamin E deficiency is similar in the human, rat and monkey. In all three species there is degeneration of the axons of the gracile and cuneate nuclei, the posterior columns and the peripheral nerves, with a selective loss of large calibre myelinated fibres (Nelson et al. 1981). These investigators concluded that the degeneration results from injury to the axonal membrane which leads to a distal and dying-back type of axonopathy with secondary demyelination.

Studies in Experimental Vitamin E Deficiency

Animal Model

The clinical observations described above pose a number of fundamental questions which can only be answered by using experimental animals. Accordingly we established a rat model of vitamin E deficiency (Goss-Sampson et al. 1988, 1990). Male Wistar rats were placed on a vitamin E-deficient diet (vitamin free casein, dextrose, stripped lard; Machlin/Draper HLR No. 814, supplied by Dyets, Pennsylvania) for up to 1 year. The control animals received the same diet to which α-tocopheryl acetate (100 mg/kg diet) was added. Up to 12 weeks of age the weights of the control and deficient rats were virtually identical (mean ± 1SD being 439.2 ± 39.1 g and 436.8 ± 40.1 g, respectively). At 16 weeks the two groups tended to diverge and by 20 weeks, the control rats were found to be significantly heavier (520.8 ± 56.1 g compared to 471.3 ± 50.9 g, $p<0.01$). From 20 to 52 weeks the controls maintained a steady growth rate whereas the growth of the vitamin E-deficient rats began to plateau at 20 weeks, and after 26 weeks there was no further increase in mean body weight. No other obvious physical differences were observed between the two groups of animals until 30–36 weeks, when the vitamin E-deficient rats began to exhibit poor coat condition, muscle wasting, hind limb weakness and impaired gait. By 50 weeks all the deficient rats had hind limb ataxia and a severe generalised muscle wasting.

There are a number of possible reasons why neural tissues may be at a particular risk from a deficiency of an antioxidant such as vitamin E. These include:

1. a reduced antioxidant status compared to other tissues;
2. the unusual structure (i.e. length) of neural cells;
3. the number of membranous structures (myelin, axolemma, intracellular membranous systems) which have high concentrations of polyunsaturated fatty acids and are, therefore, likely to be susceptible to lipid peroxidation; and
4. the very slow or absent turnover of neural cells. We have begun to examine some of these factors, as discussed below.

Electrophysiological Studies

A deficiency of vitamin E in man results in characteristic electrophysiological abnormalities (Muller and Goss-Sampson 1990). Thus the measurement of similar electrophysiological modalities in the vitamin E-deficient rat provided an objective means of validating this model.

A comparison of the results observed in vitamin E-deficient humans and data obtained in the rat model is shown in Table 1. In humans central conduction was more severely affected than peripheral conduction, and abnormalities of sensory function were more frequent than motor (Muller and Goss-Sampson 1990). It is of interest that, despite a delay in central conduction, the brainstem auditory evoked potentials were consistently normal (Muller and Goss-Sampson 1990).

Table 1. Electrophysiological abnormalities in vitamin E deficiency

Modality	Humans	Rats
Reference	Muller and Goss-Sampson (1990)	Goss-Sampson et al. (1988, 1990, 1991)
Central conduction	Yes	Yes
Peripheral conduction	Rare	Late
Peripheral motor conduction	No	No
Auditory evoked potentials	No	No
Electroretinography	Yes	Yes
Visual evoked potentials	Yes	Yes
Electromyography	Yes	Yes

The results in the rat were essentially similar (Goss-Sampson et al. 1990). There was a significant slowing in central conduction after 8 months of deficiency, whereas significant changes in peripheral conduction only became detectable after 11 months. No changes in brainstem auditory evoked potentials and peripheral motor conduction were found.

With regard to visual function, abnormalities of the electroretinogram and visual evoked response have been reported in vitamin E-deficient humans, with the former appearing to precede the latter (Muller and Goss-Sampson 1990). After 10 months of deficiency in the rat (earlier-time points have not been examined), significant changes in both electroretinography and visual evoked potentials were observed (Goss-Sampson et al. 1991).

Abnormalities of the electromyogram (fibrillation potentials and polyphasic groups) have been reported in a number of studies in man (Muller and Goss-Sampson 1990) and are generally considered to be consistent with chronic partial denervation. Similar abnormalities of the electromyogram were found in the vitamin E-deficient rat (Goss-Sampson et al. 1988).

The similarities in the electrophysiological findings in vitamin E-deficient humans and rats confirm the validity of the rat model. The observations are consistent with the clinical and neuropathological findings described above. They are what would be expected from a distal dying-back axonal neuropathy which is more severe in the posterior columns than in peripheral nerves.

Antioxidant Status

The vitamin E and antioxidant status of normal rats (Goss-Sampson and Muller 1987) and longitudinal studies during vitamin E deficiency (Goss-Sampson et al. 1988) have been reported previously. In summary α-tocopherol concentrations in the brain were highest in the cortex and lowest in the cerebellum ($P<0.02$). The activity of glutathione peroxidase tended to show an inverse relationship to the α-tocopherol concentration, whereas activities of superoxide dismutase were evenly distributed throughout the brain. Concentrations of α-tocopherol increased in the lower spinal cord (region of the lumbar/sacral outflow) and were evenly distributed along the sciatic and tibial nerves.

In rats on a vitamin E-deficient diet, the concentration of the vitamin declined rapidly in liver, adipose tissue and plasma and was undetectable by 16–24 weeks (see Fig. 1). In all regions of the nervous system (brain, spinal cord and nerve), the rate of decline was slower than in other tissues, and by 52 weeks of deficiency, concentrations were still approximately 5 % of the control values (Fig. 1). This capacity of nervous tissue to retain α-tocopherol may account for the delay in onset of symptoms noted in human subjects. After 52 weeks of deficiency the concentrations of α-tocopherol were similar in all the brain regions (20–24 μg/g lipid), with no region exhibiting selective retention or loss. The concentration of α-tocopherol in the various regions of the spinal cord was approximately twice that in the brain (43–50 μg/g lipid). There was no selective decrease in the various nerve regions, with concentrations varying between 23–47 μg/g lipid.

The observation that neural tissues preferentially conserve vitamin E has been found by others. Vatessery et al. have reported that neural tissues of vitamin E-deficient mice preferentially conserve vitamin E (Vatassery et al. 1984), and Ingold et al. (1987), using deuterated tocopherols, found that the half-life of tocopherol was greater in neural than in non-neural tissues. For example, the half-life of α-tocopherol in liver and plasma was approximately 10 days, whereas in brain and spinal cord it was 29 and 76 days, respectively.

The activities of superoxide dismutase and glutathione peroxidase have been estimated over a 52-week period in the various brain regions of control and deficient animals (Goss-Sampson et al. 1988). At no time points were there any significant differences between the two groups of animals in the activities of either enzyme. This indicates that these two aqueous antioxidant systems do not compensate for the reduced vitamin E concentrations.

Lipid Peroxidation

In vitro and probably in vivo, vitamin E acts as a lipid-soluble secondary antioxidant and would be expected to protect neural membranes with their high concentrations of polyunsaturated fatty acids from lipid peroxidation. We have, therefore, investigated lipid peroxidation in neural and other tissues, and also in different neural fractions during the course of experimental vitamin E deficiency. The concentrations of thiobarbituric acid reactive substances (TBARS), free malondialdehyde (MDA) and the aliphatic aldehydes (propanal, butanal, pentanal, hexanal and 4-hydroxynonenal) have been measured as indices of lipid peroxidation. All tissues studied from 1-year-old vitamin E-deficient rats showed an increase in the concentration of TBARS compared to controls. The increase was greater (approximately two- to four fold) in peripheral tissues (liver, muscle and heart) than in neural tissues (brain and spinal cord), where increases were less than 35 %. All the parameters studied (with the exception of butanal) showed a significant increase in the cerebral cortex of 52-week vitamin E-deficient rats compared to controls, with the aliphatic aldehydes showing the greatest increases (approximately six fold).

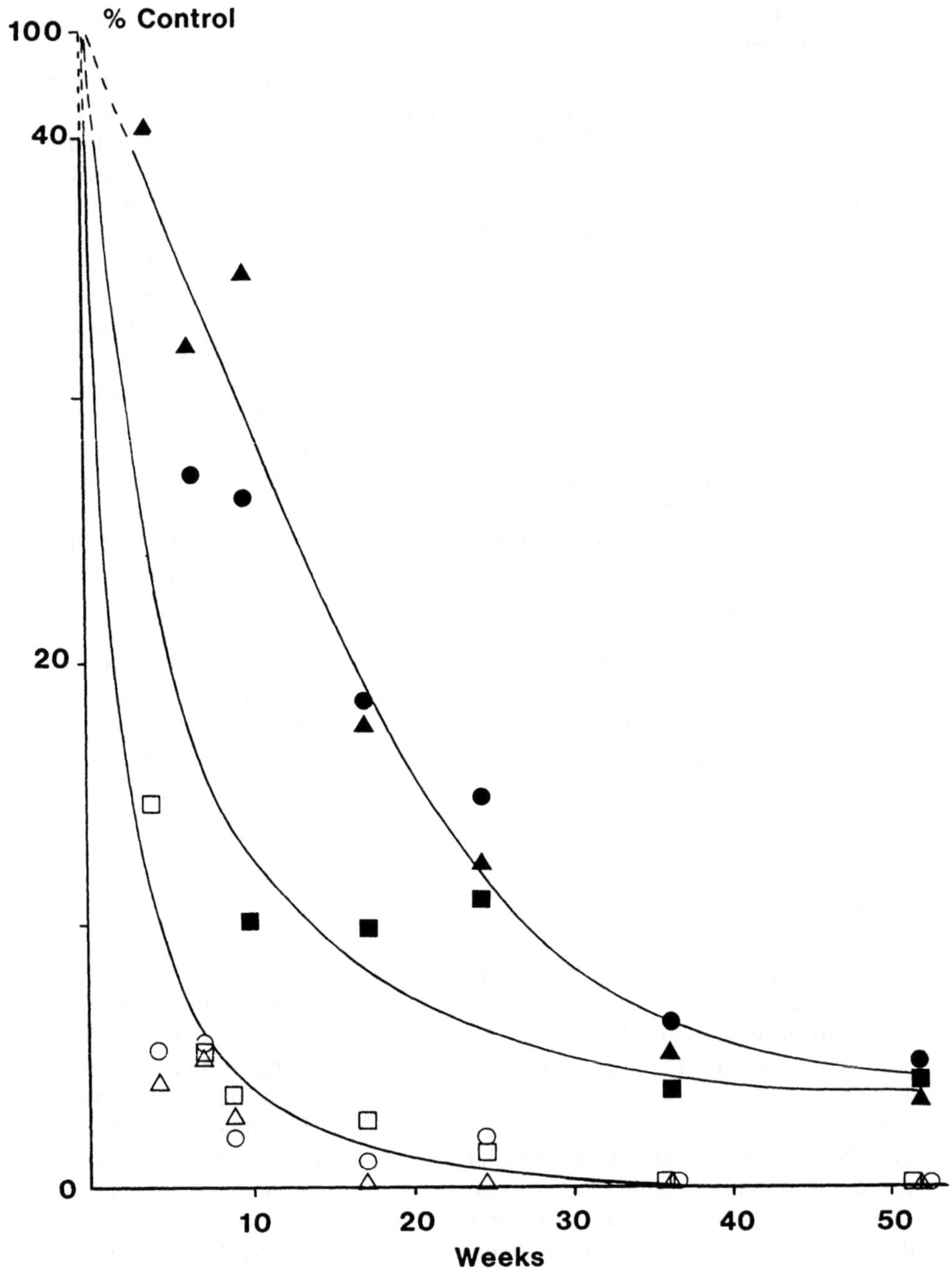

Fig. 1. Mean concentrations of α-tocopherol (μg/g lipid) in tissues from vitamin E-deficient rats (expressed as a percentage of control values) from weaning to 52 weeks (n=3). ▲ brain; ● cord; ■ nerve; △ serum; ○ liver; □ adipose tissue

In addition to studying whole neural tissues, myelinated nerves from rat brainstem were separated by density gradient ultracentrifugation (De Vries et al. 1981) into three membrane fractions (myelin, axolemma rich fraction, and a fraction containing membrane organelles and intracellular membranes). Whole neural tissues and the neural fractions from vitamin E-deficient and control animals were stressed in vitro with a free radical-generating system ($CuSO_4/H_2O_2$) and the subsequent lipid peroxidation was assessed by the measurement of free MDA. In summary the following observations were made:

1. The deficient neural fractions did not exhibit a lag phase whereas, with the exception of myelin, the control fractions showed a lag phase of at least 10 min.
2. The neural tissues and membranes showed the following order of susceptibility to lipid peroxidation: brain >> spinal cord > nerve, and intracellular membrane fraction >> axolemma enriched fraction > whole homogenate> myelin. The order of susceptibility of the neural fractions is of interest as it agrees with the known neuropathology associated with vitamin E deficiency, where the axon is initially affected, resulting in a primary axonopathy (Nelson et al. 1981). The sensitivity of the intracellular membrane fraction to free radical attack could also explain the abnormalities of axonal transport seen in vitamin E-deficient rats, as discussed below.

Axonal Transport

Neural cells have complex transport systems to carry materials and organelles from the cell body of the neuron down the length of its axon (anterograde transport) and to return them back to the cell body for reprocessing (retrograde transport). As impaired axonal transport is known to be associated with axonal degeneration, fast anterograde and retrograde transport of acetylcholinesterase was measured in vitamin E-deficient and control rats. This transport was determined in the sciatic nerve, using a double ligation/accumulation technique (Oikarinen and Kalimo 1984). The results at 52 weeks are shown in Fig. 2. The deficient animals showed a 25 % reduction in anterograde transport ($P<0.02$) and a 20 % reduction in retrograde transport ($P<0.05$). No significant differences in either anterograde or retrograde transport were found at earlier time points. The reduced axonal transport, together with ultrastructural studies of sensory nerve endings which show an accumulation of normal and abnormal organelles, multivesicular bodies, and tubulovesicular structures (Southam et al. 1991), suggest there may be a defect in "turnaround" at the axon terminal.

Conclusions

On the basis of the studies described above it is possible to postulate a sequence of events to explain the development of the neurological syndrome associated with vitamin E deficiency (Fig. 3). A deficiency of vitamin E would be expected

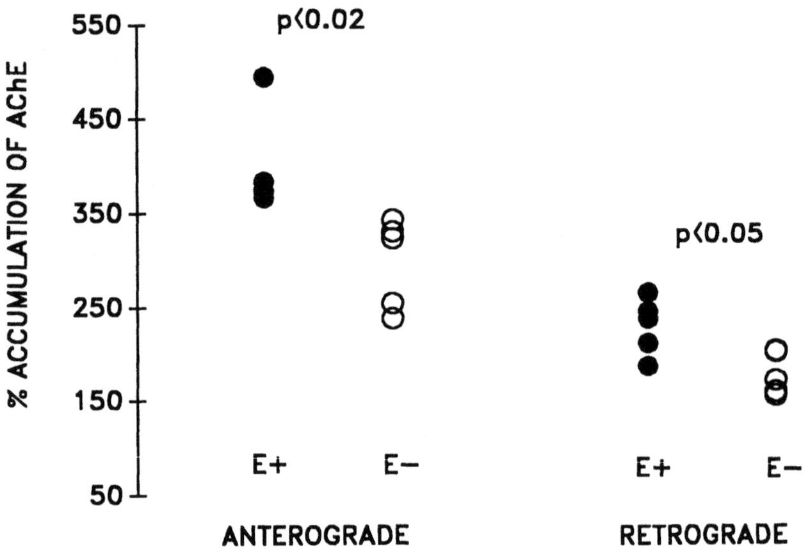

Fig. 2. Fast anterograde and retrograde axonal transport of acetylcholinesterase (AchE) in sciatic nerves of control (●) and vitamin E-deficient (○) rats after 52 weeks

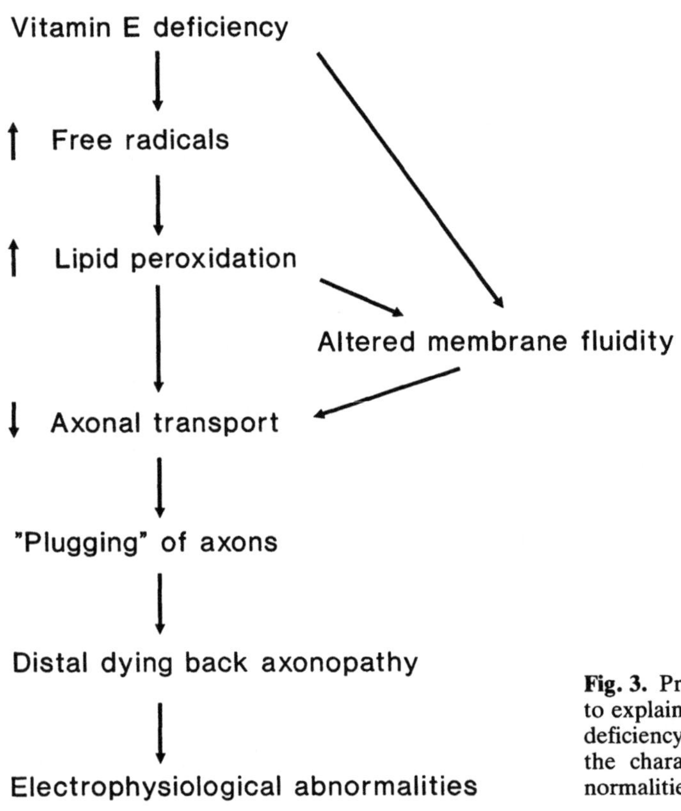

Fig. 3. Proposed sequence of events to explain how a severe and chronic deficiency of vitamin E may result in the characteristic neurological abnormalities

to result in increased concentrations of oxygen derived free radicals and increased lipid peroxidation. Membrane organelles and the intracellular membranous systems appear to be particularly susceptible to attack by oxygen derived free radicals. If, for example, there is a disturbance of the axonal mitochondria, this might be expected to lead to abnormalities in fast anterograde and retrograde transport, as both processes are energy dependent. This in turn would result in the accumulation of organelles and other material at the nerve terminal which would "plug off" the axons. This would tend to isolate the nerve endings from the cell body, causing them to degenerate and thereby resulting in the primary axonopathy. This process could then spread back along the nerve in a dying-back manner. The axonopathy could then lead to a secondary demyelination resulting in the characteristic neuropathological and electrophysiological abnormalities associated with vitamin E deficiency.

In addition to its role as an antioxidant, α-tocopherol may also have a structural role in membranes (Diplock and Lucy 1973). Changes in membrane fluidity as a result of decreased concentrations of α-tocopherol and/or increased lipid peroxidation may, therefore, also result in altered axonal transport and contribute to the neural abnormalities. Membrane fluidity is currently being investigated in membranes from vitamin E-deficient and control rats by fluorescence polarography.

Acknowledgements. Financial support from the Friedreich's Ataxia Group, F. Hoffmann-La Roche and Company Limited, the Leverhulme Trust and the Merchant Taylor's Company is gratefully acknowledged.

References

Alvarez F, Landrieu P, Laget P, Lemonnier F, Odievre M, Alagille D (1983) Nervous and ocular disorders in children with cholestasis and vitamin A and E deficiences. Hepatology 3: 410–414

Azizi E, Zaidman JL, Eschar J, Szeinberg A (1978) Abetalipoproteinaemia treated with parenteral and oral vitamins A and E and with medium chain triglycerides. Acta Pediatr Scand 67: 797–801

Burton GW, Joyce A, Ingold KU (1983) Is vitamin E the only lipid-soluble, chain-breaking antioxidant in human blood plasma and erythrocyte membranes? Arch Biochem Biophys 221: 281–290

DeVries GH, Zetusky WJ, Zmachinski C, Calabrese VP (1981) Lipid composition of axolemma-enriched fractions from human brains. J Lipid Res 22: 208–216

Diplock AT, Lucy JA (1973) The biochemical modes of action of vitamin E and selenium: a hypothesis. FEBS Lett 29: 205–21

Elias E, Muller DPR, Scott J (1981) Association of spinocerebellar disorders with cystic fibrosis or chronic childhood cholestasis and very low serum vitamin E. Lancet ii: 1319–1321

Evans HM, Bishop KS (1922) On the existence of a hitherto unrecognised dietary factor essential for reproduction. Science 56: 650–651

Goss-Sampson MA, Muller DPR (1987) Studies on the neurobiology of vitamin E (alpha-tocopherol) and some other antioxidant systems in the rat. Neurpathol Appl Neurobiol 13: 289–296

Goss-Sampson MA, MacEvilly CJ, Muller DPR (1988) Longitudinal studies of vitamin E and other antioxidant systems, and neurological function in the vitamin E deficient rat. J Neurol Sci 87: 25–35

Goss-Sampson MA, Kriss A, Muller DPR (1990) A longitudinal study of somatosensory, brainstem auditory and peripheral sensory-motor conduction during vitamin E deficiency in the rat. J Neurol Sci 100: 79–84

Goss-Sampson MA, Muller DPR, Kriss A (1991) Abnormalities of the electroretinogram and visual evoked potential in vitamin E deficient rats. Exp Eye Res 53: 623–627

Guggenheim MA, Ringel SP, Silverman A, Grabert BE (1982) Progressive neuromuscular disease in children with chronic cholestasis and vitamin E deficiency: diagnosis and treatment with alpha-tocopherol. J Pediatr 100: 51–58

Halliwell B, Gutteridge JMC (1985) Oxygen radicals and the nervous system. Trends Neurosci 8: 22–26

Halliwell B, Gutteridge JMC (1989) Free radicals in biology and medicine, 2nd edn. Oxford University Press, Oxford

Harding AE, Muller DPR, Thomas PK, Willison HJ (1982) Spinocerebellar degeneration secondary to chronic intestinal malabsorption: a vitamin E deficiency syndrome. Ann Neurol 12: 419–424

Harries JT, Muller DPR (1971) Absorption of vitamin E in children with biliary obstruction. Gut 12: 579–584

Hegele A, Angel A (1985) Arrest of neuropathy and myopathy in abetalipoproteinemia with high-dose vitamin E therapy. Can Med Assoc J 132: 41–44

Herbert PN, Gotto AM, Fredrickson DS (1978) Familial lipoprotein deficiency. In: Stanbury JB, Wyngaarden JB, Fredrickson DS (eds) The metabolic basis of inherited disease, 4th edn. McGraw Hill, New York, pp 544–588

Howard L, Oveson L, Satya-Murti S, Chu R (1982) Reversible neurological symptoms caused by vitamin E deficiency in a patient with short bowel syndrome. Am J Clin Nutr 36: 1243–1249

Ingold KU, Burton GW, Foster DO, Hughes L, Lindsay DA, Webb A (1987) Biokinetics of and discrimination between RRR- and SRR-α-tocopherols in the male rat. Lipids 22: 163–172

Kane JP, Havel RJ (1989) Disorders of the biogenesis and secretion of lipoproteins containing the B apolipoproteins. In: Scriver CR, Beaudet AL, Sly WS, Valle D (eds) The metabolic basis of inherited disease, 6th edn. McGraw Hill, New York, pp 1139–1164

Kayden HJ, Silber R, Kossmann CE (1965) The role of vitamin E deficiency in the abnormal autohemolysis of acanthocytosis. Trans Assoc Am Physicians 78: 334–342

Miller RG, Davis CJF, Illingworth DR, Bradley W (1980) The neuropathy of abetalipoproteinemia. Neuropathy 30: 1286–1289

Muller DPR, Lloyd JK, Bird AC (1977) Long-term management of abetalipoproteinaemia. Arch Dis Child 52: 209–214

Muller DPR, Lloyd JK (1982) Effect of large oral doses of vitamin E on the neurological sequelae of patients with abetalipoproteinemia. Ann NY Acad Sci 393: 133–144

Muller DPR, Goss-Sampson MA (1990) Neurochemical, neurophysiological and neuropathological studies in vitamin E deficiency. Crit Rev Neurobiol 5: 239–263

Muller DPR, Harries JT, Lloyd JK (1974) The relative importance of the factors involved in the absorption of vitamin E in children. Gut 15: 966–971

Nelson JS, Fitch CD, Fischer VW, Broun GO, Chou AC (1981) Progressive neuropathologic lesions in vitamin E deficient Rhesus monkeys. J Neuropathol Exp Neurol 40: 166–186

Oikarinen R, Kalimo H (1984) Acetylcholinesterase activity and its fast axonal transport in rabbit sciatic nerves during the recovery phase of experimental allergic neuritis. Neuropathol Appl Neurobiol 10: 163–171

Pappenheimer AM, Goettsch M (1931) A cerebellar disorder in chicks apparently of nutritional origin. J Exp Med 53: 11–26

Rosenblum JL, Keating JP, Prensky AL, Nelson JS (1981) A progressive neurologic syndrome in children with chronic liver disease. N Engl J Med 304: 503–508

Sokol RJ, Heubi JE, Iannaccone S, Bove KE, Ballistreri WF (1983) Mechanism causing vitamin E deficiency during chronic childhood cholestasis. Gastroenterology 85: 1172–1182

Sokol RJ, Guggenheim MA, Iannaccone ST, Barkhaus PE, Miller C, Silverman A, Ballistreri WF, Heubi JE (1985) Improved neurologic function after long term correction of vitamin E deficiency in children with chronic cholestasis. N Engl J Med 313: 1580–1586

Sokol RJ, Kayden HJ, Bettis DB, Traber MG, Neville H, Ringel S, Wilson WB, Stumpf DA (1988) Isolated vitamin E deficiency in the absence of fat malabsorption – familial and sporadic cases: characterisation and investigation of causes. J Lab Clin Med 111: 548–559

Southam E, Thomas PK, King RHM, Goss-Sampson MA, Muller DPR (1991) Experimental vitamin E deficiency: morphological and functional evidence of abnormal axonal transport secondary to free radical damage. Brain 114: 915–936

Traber MJ, Sokol RJ, Burton GW, Ingold KU, Papas AM, Huffaker JE, Kayden HJ (1990) Impaired ability of patients with familial isolated vitamin E deficiency to incorporate α-tocopherol into lipoproteins secreted by the liver. J Clin Invest 85: 397–407

Vatassery GT, Angerhofer CK, Peterson FJ (1984) Vitamin E concentrations in the brains and some selected peripheral tissues of selenium-deficient and vitamin E-deficient mice. J Neurochem 42: 554–558

Wasserman RH, Taylor AN (1972) Metabolic roles of fat-soluble vitamins D, E and K. Annu Rev Biochem 41: 179–202

Willison HJ, Muller DPR, Matthews S, Jones S, Kriss A, Stead RJ, Hodson ME, Harding AE (1985) A study of the relationship between neurological function and serum vitamin E concentrations in patients with cystic fibrosis. J Neurol Neurosurg Psychiatry 48: 1097–1102

Essential Fatty Acids and Alpha-Tocopherol Supplementation in Tardive Dyskinesia

K.S. Vaddadi

Summary

TD is a disorder of movement that can emerge during the course of treatment with neuroleptics or during their withdrawl or spontaneously. Its cause is not well understood but is considered to be due to DA receptor supersensitivity, due to chronic neuroleptic blockade of DA receptors. There is no definitive treatment. Free radicals have been implicated in the aetiology of TD. Alpha-tocopherol and ascorbic acid are known free radical scavengers (antioxidants). Clinical trials in patients with TD have shown reductions in the severity of TD on administration of 1200 IU daily of alpha-tocopherol (vitamin E). Vitamin E and ascorbid acid (vitamin C) in vitro studies have an overall additive effect as an antioxidant.

EFA are important constituents of every cell in the body. They have an important role to play in the structure and function of a cell membrane. Animal studies have demonstrated antidyskinetic properties of dihomo-gamma-linolenic (DGLA) acid. Some human studies have shown antidyskinetic properties of gamma-linolenic acid. Long-term studies using a combination of EFA and antioxidants as a preventative measure in patients taking neuroleptics need to be undertaken.

Introduction

Schizophrenia is a major psychotic illness. Two principal forms of the illness have been described (Crow 1989). Type I is characterised by the presence of so-called positive symptoms – hallucinations, delusions, disordered thinking. Type II schizophrenia has predominant negative symptoms (e.g., affective flattening, poverty of speech), intellectual impairment, behavioral deterioration and dyskinesias. Although the clinical correlates of structural changes in the schizophrenic brain have been variable between studies, when present they tend to have been among the features of Type II syndrome in schizophrenia (Crow 1989). Neuropathological changes in schizophrenia suggest that the medial temporal lobes are preferentially affected and that these changes may well take place before the development of symptoms (Roberts and Crow 1987). However, patients without negative symptoms also show abnormal involuntary movements. Essential fatty acids and antioxidants like alpha-tocopherol (vitamin E) are important in maintaining the neuroarchitecture and functioning of the nerve cell. Essential fatty acids (EFAs) are required for normal dopaminergic receptor functions.

Polyunsaturated fatty acids are susceptible to lipid peroxidation by free radicals generated during biological processes. Antioxidants act as scavengers of the free radicals and prevent tissue damage. It is likely that EFAs and antioxidants play an important role in certain neuropsychiatric conditions, like schizophrenia, tardive dyskinesia and Parkinson's disease. The work done so far suggests that EFAs, and the co-factors essential in their metabolism, and alpha-tocopherol on its own have an antidyskinetic effect in both animal and human studies. In addition they have a distinct positive effect in improving mental state in schizophrenic patients. Some benefit has also been observed from gamma-linolenic acid in improving memory score and negative symptoms in schizophrenia. However, the studies described in this chapter will focus on the role of alpha-tocopherol and essential fatty acids in alleviating dyskinesia.

Tardive Dyskinesia

Abnormal movements like mannerisms and stereotypies had been reported long before the use of neuroleptics in schizophrenia. Movements like grimacing and irregular tongue movements were observed by Kraepelin (1907) and Bleuler (1950). Kraepelin believed that the abnormal movements were an integral part of the biological basis of schizophrenia. Spontaneous involuntary movements have been found to be a feature of chronic schizophrenia, unmodified by neuroleptics (Owens-Cunningham et al. 1982). The use of neuroleptics and their effects on the extrapyramidal system have confounded the aetiology of these abnormal movements. At the same time, tardive dyskinesia (TD) has been reported in non-psychiatric populations exposed to drugs that are dopamine antagonists (metoclopramide); Wiholm et al. 1984; Breitbart 1986).

Tardive dyskinesia (TD) is characterised by involuntary, repetitive, purposeless and hyperkinetic movements of the body. Abnormal movements in the areas around the mouth, tongue and eyes are common (tongue protrusion, worm-like movements of tongue, sucking, pouting, etc.). Other areas affected can be the upper and lower limbs and axial muscles. Breathing and swallowing dyskinesias have also been reported (Casey 1987). The commonly used diagnostic criteria are:

1. presence of abnormal involuntary stereotyped movements
2. exposure to neuroleptics for a period of at least three months
3. the dyskinesias disappear during sleep
4. voluntary effort and activity of the muscles in the affected areas can temporarily suppress the movements
5. stress and movements in distant parts of the body will aggravate the dyskinesias
6. they are not helped by anticholinergics
7. other neurological disorders likely to cause similar movement have been excluded (from Jeste et al. 1979)

A prevalence of 20 % has been reported among patients receiving long-term neuroleptics (Marsden and Fahn 1987). Schizophrenic patients in particular do not actively complain or seek treatment for their TD symptoms, although staff and relatives do. The most commonly accepted aetiological explanation for TD is that it results from denervation hypersensitivity, caused by chronic dopaminergic receptor blockade.

Antioxidants

A major defence of the cell against free radicals and other oxidative damage includes the antioxidant vitamins, like alpha-tocopherol and ascorbate, and enzymes like catalase, superoxide dismutase and compounds like glutathione (Halliwell and Gutteridge 1985). Two types of antioxidants are known: the preventive oxidants, such as catalase and peroxidase, which reduce the formation of free radicals by decomposing peroxide without the generation of free radicals; and the chain-breaking antioxidants, which scavenge free radicals to terminate free radical reactions and prevent chain propagation reactions (Etsuo 1987).

Alpha-Tocopherol and Ascorbic Acid

Tocopherols occur mainly in the unesterified form in a variety of plants such as nuts, seeds, oils, fruits and grasses. The family of tocopherols alpha, beta, gamma and delta are dependent upon the aromatic methyl (CH_3) groups being in either the 5, 7 or 8 position of the ring. When the form of the vitamin is d-alpha tocopherol, it is from natural sources, while if the vitamin is dl-alpha tocopherol, then it is synthetic vitamin E. Its absorption depends on the dose and the vehicle of administration. Polyunsaturated fatty acids decrease the absorption of alpha-tocopherol. Evidence suggests that it is more effectively absorbed in the aqueous form (Regtop 1983). There is evidence of increased triglycerides in the serum when alpha-tocopherol is administered, hence patients taking alpha-tocopheron need triglyceride level monitoring (Regtop 1983). Values of 0.8 mg of tocopherols per gram of total lipids in the serum of plasma of adults and 0.6 mg for children have been suggested as limits for assessment of vitamin E adequacy in humans (Howitt et al. 1972; Farrell et al. 1978).

In humans severe and prolonged vitamin E deficiency results in a progressive neurological syndrome characterised by ataxia of limbs, abnormalities of gait, muscle weakness, areflexia, proprioceptive loss, opthalmoplegia and retinal pigmentation (Muller et al. 1983). Tocopherols act as antioxidants in the phospholipid liposomal membrane. They suppress the oxidation of phospholipids induced by free radicals. They scavenge both oxygen radicals attacking the membrane from outside and chain-carrying lipid peroxyl radicals within the membrane and prevent cell damage.

Ascorbic acid (vitamin C) is required for human health and, like EFA, cannot be synthesized by humans. Hence it has to be provided in the diet. It is important for the optimal action of a number of enzymes (Levin 1986). Ascorbate is a powerful reducing agent and chain-breaking water soluble antioxidant and can scavenge oxygen radicals in aqueous phase (Etsuo 1987). The effect of ascorbate on peroxidation is known to depend on its concentration as well as on the oxidant system. In in vitro studies it can be a pro- or antioxidant. However, in the presence of both alpha-tocopherol and ascorbate, the overall effect is protective and exceeds the sum of individual effects against peroxidation.

Ascorbid acid regenerates vitamin E, by reducing vitamin E radicals formed when vitamin E scavenges the oxygen radicals (Etsuo 1987). It appears, therefore, that the synergistic action between vitamins E and C may have a considerable therapeutic potential in TD, by reducing the damage to the neuronal cells from free radicals.

Free radicals as an aetiological factor in TD and clinical studies with alpha-tocopherol in its therapy.

The term "radical" is often used loosely in chemistry. A broad definition of "free radical" given by Halliwell and Gutteridge (1985) is as follows: A free radical is any species capable of independent existence that contains one or more unpaired electrons. (An unpaired electron is one that occupies an atomic or molecular orbit by itself).

The presence of one or more unpaired electrons makes the species highly reactive. The free radical hypothesis of tardive dyskinesia suggests that structural damage to catecholaminergic as well as other neurotransmitter systems may be due to free radicals generated due to the chronic use of neuroleptics (Cadet et al. 1986). The cell destruction in the region of the basal ganglia especially the striatum and substantia nigra may be important in the aetiology of TD. Similar mechanisms of free radical production due to increased turnover of dopamine as a compensatory mechanism, for example in idiopathic Parkinson's disease which is due to the loss of dopamine (DA)-containing neurons, have also been suggested (Hornykiewicz 1982; Fahn 1989). It is therefore possible that, by preventing neuronal peroxidation damage and by stabilising the cell membrane (Diplock 1983) alpha-tocopherol may have a therapeutic potential in the treatment of TD, (Lohr et al. 1988a).

In an animal model study, Lohr et al. (1988b) treated rats with the neurotoxin iminofipropionitrile (IDPN). This causes alpha-tocopherol deficiency-like axonal damage and irreversible movement disorder. Animals receiving 2 gms/kg of alph-tocopherol concurrently with 100 mg/kg IDPN for 10 days demonstrated a significant reduction in IDPN-induced vertical head movements compared to the animals receiving IDPN alone. When they were first treated with IDPN and then given vitamin E, a significant reduction in the severity of IDPN-induced dyskinesia was produced, although in both experiments locomotor activity was unchanged. The authors suggested the possible involvement of free radical formation in the neurotoxicity of IDPN.

In another study conducted by Lohr and others (1987; 1988 a) 15 patients with a diagnosis of schizophrenia or schizo-affective disorder with persistent TD for at least one year consented to take alpha-tocopherol or placebo in a randomised blind crossover trial. A modified abnormal involuntary movements scale (AIMS) score and a modified version of the Simpson-Angus scale for extrapyramidal side effects were used (Simpson and Angus 1970). The initial AIMS score was 13.9 (Lohr et al. 1987) the mean reduction with active alpha-tocopherol treatment in AIMS score was 43 % and seven patients showed greater than 50 % improvement in their dyskinesia score. A subset of 12 patients completed a second placebo and active treatment phase which showed a significant overall reduction in AIMS score with active alpha-tocopherol treatment and not with the placebo treatment. There was no significant difference between the two placebo periods in patients who completed both the phases of the clinical trial (Fig. 1).

When mental state scores were examined using the Brief Psychiatric Rating Scale (BPRS) scores, alpha-tocopherol had a significant effect in reducing anxiety and depression subscale scores, with a nonsignificant trend for a decrease in total BPRS scores. The patients who showed greater than 50 % improvement in their AIMS scores had shorter duration of TD and later age of onset, compared to the group of patients with less than 50 % improvement.

The two groups were similar on other clinical variables. These results are extremely encouraging but, as suggested by Lohr et al. (1988 a) the drug interactional effect between alpha-tocopherol and neuroleptics is unclear. However, it is unlikely that one would expect benefit from alpha-tocopherol in advanced TD cases, as the neuronal damage already exists. Hence this treatment

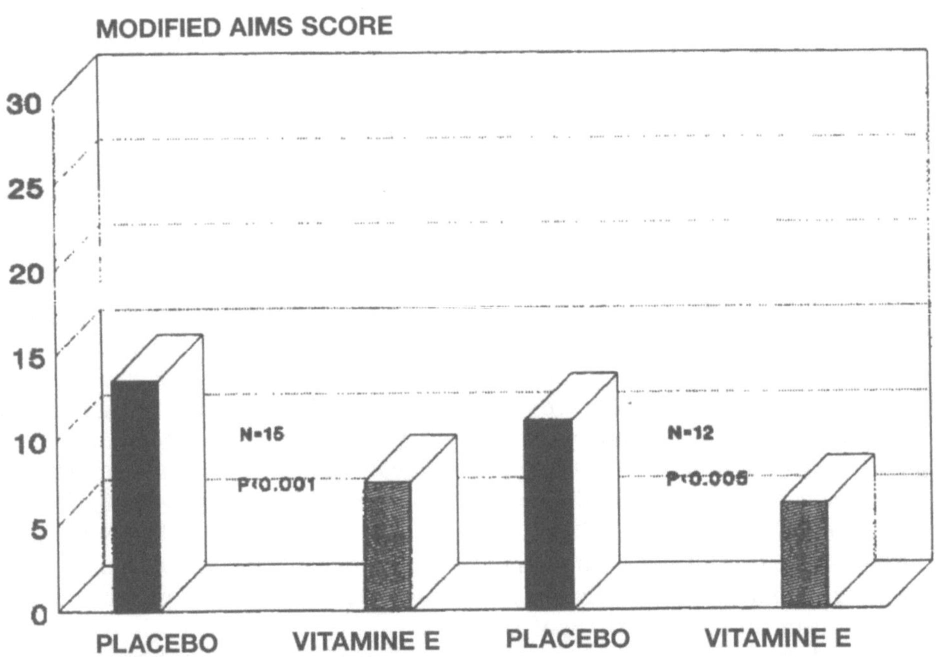

Fig. 1. Vitamin E in tardive dyskinesia (From Lohr et al. 1987)

may turn out to have a role in prevention of TD in patients who are likely to receive neuroleptics (depot preparations) for longer periods.

In addition, TD has a fluctuating clinical course and bears some relationship with mood state (Cutler et al. 1981). Alpha-tocopherol appears to have an effect on mood-related items of BPRS. When patients with a diagnosis of schizo-affective disorder are also included in the clinical studies of TD, changes in the patient's mood state induced by alpha-tocopherol may confound the results, due to the relationship between mood state and TD.

Furthermore, patients with schizo-affective disorder may well be receiving lithium carbonate, as polypharmacy is not uncommon in psychiatric practice. Lithium has an effect in the mobilitsation of dihomo-gamma-linolenic acid (DGLA) from cell membranes (Horribin 1978). Essential fatty acids (EFAs) have a regulatory action on ion channels (Ordway et al. 1991). They are precursors of prostaglandins (PGs). PG have a neuromodulatory effect, in addition to an effect on DA receptor function. All these actions and interactions are likely to produce alterations in TD scores and complicate the true measurement of effects of an experimental drug.

Therefore, diagnostic status, current mood state and mood altering drugs' effects must be taken into consideration in rating TD patients and interpreting the data. In an ideal situation one would like to have control over dietary factors that may have some influence on neuronal functions, such as antioxidants, smoking, and EFA intake. However, this may not be practical, as a large proportion of TD patients are chronic schizophrenics and co-operation is not ideal.

In another study, by Elkashef et al. (1990) eight patients (schizophrenics and schizo-affectives) with TD (mean age 56.6 ± 12.0 years) who had taken neuroleptics for at least 15 years and with a mean duration of TD of 3.8 ± 2.5 years were included in a clinical trial. They were given alpha-tocopherol or placebo for a period of four weeks in a double-blind fashion with a crossover for a further four weeks. Five of the eight patients showed reductions of 30 % or more on AIMS score. Two patients with bucco-lingual movements or with dystonia improved the most. There was no change in their mental state scores. The dosage schedule of alpha-tocopherol in both studies (Lohr et al. 1987; Elkashef et al. 1990) had been a gradual increase to 1200 IU daily. However, the sample sizes were not large. It is essential to observe the effect of alpha-tocopherol on TD over much longer periods to see if the antidyskinetic effect is a sustained one, as the nature of TD itself is varaible.

A recently presented paper by Kontsevoy et al. (1991), describing an open study on 20 elderly patients with TD, reported a 50 % decrease in a four-week period in 50 % of the patients and a 43.7 % improvement in TD scores (AIMS scores) for the entire group. However, this study was uncontrolled and no placebo group was included.

Shriqui et al. (1991), in a recent double-blind crossover study, compared alpha-tocopherol with placebo in a group of 33 subjects with TD. Twenty-seven completed the trial (mean age 42.9 ± 12.6 years) 22 had schizophrenia and five had a bipolar disorder. In a six-week treatment period, using alpha-tocopherol up to 1200 IU daily, no significant clincial effects were observed.

Although negative results with the use of alpha-tocopherol are emerging, long-term studies using optimum dosage and larger homogenous samples are needed to evaluate the results.

Essential Fatty Acids and TD

Essential Fatty Acids and their metabolic pathways

Essential fatty acids (EFA) are polyunsaturated fatty acids and must be taken in the diet as the human body cannot make them (Holman 1966; Mead and Fulco 1976). There are two types of EFA, the N-6 series derived from linoleic acid (18:2N-6) and the N-3 series derived from alpha-linolenic acid (18:3N-3; Holman 1966). There is another sequence of fatty acids in which oleic acid (18:1N-9) corresponds to linoleic and alpha-linolenic acid of EFA series. The N-9 series fatty acids are not essential, since oleic acid can be readily manufactured in mammals from simple precursors.

However, the N-3, N-6 and N-9 series of unsaturated fatty acids are all metabolised by the same sequence of enzymes with alternating desaturases which remove two hydrogens and elongases which add two carbons. The enzyme sequence seems to metabolise the fatty acids with the following sequence of preference: N-3 first, N-6 second and N-9 third. These three fatty acids compete with each other in their metabolism, especially at the desaturation steps. For EFA activity all the double bonds must be in cis configuration. Alpha-linolenic acid does not prevent all the symptoms caused by a fat-free diet. EFAs are rapidly oxidised in the presence of oxygen with the formation of peroxides and toxic free radicals, and they are protected by antioxidants like alpha-tocopherol.

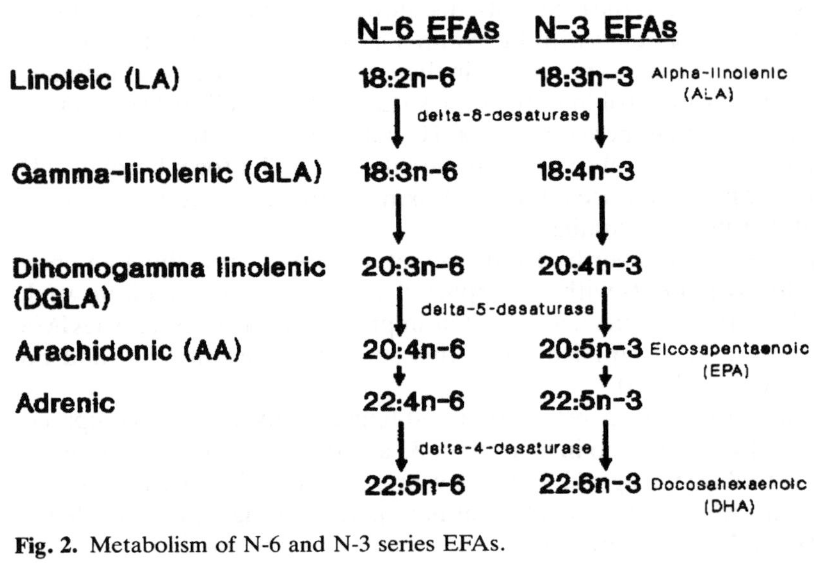

Fig. 2. Metabolism of N-6 and N-3 series EFAs.

Alpha-tocopherol also inhibits lipo-oxygenase and it is possible that its absence may convert EFA to large amounts of toxic lipo-oxygenase products. Phenothiazines prescribed for psychiatric conditions form free radical intermediates during their metabolism. These could cause damage to nervous tissue that contains EFAs. Elevated levels of the products of peroxidation in the CSF have been found in patients receiving phenothiazines (Pall et al. 1987).

Functions of EFAs

The EFAs are important constituents of biological membranes. Lipids in the form of EFAs constitute approximately 20 % of the solid matter in the brain (Horrobin 1982). They have two major roles: forming the structure of the cell membrane and in the biosynthesis of a variety of short-lived derivatives like eicosanoids, prostaglandins (PG) and leukotrienes. They confer on membranes properties of fluidity and flexibility and modulate the behaviour of membrane-bound proteins such as receptors, ATPases and ion channels. A number of studies have now shown that the non-esterified or "free" form of fatty acids regulates ion channels (Horrobin and Manku 1990; Ordway et al. 1991). These actions on ion channels can be direct or through the arachidonic acid metabolites, which then influence the ion channels. Changes in the cellular content of non-esterified or "free" fatty acids might have a profound effect on cell function. EFAs are precursors of many short-lived substances, like PG. PGs modulate nerve conduction, neurotransmitter release and post-synaptic transmitter actions (Horrobin 1982). They are found in the brain and their levels change rapidly following changes of dietary EFA intake (Hassam et al. 1980). Unlike the situation in other tissues, in the brain, almost all the fatty acids present are 6 desaturated (Horrobin 1990 a).

Due to the profound influence of EFAs and PGs on brain function, it is most likely that they have some role to play in motor function.

EFA as a treatment strategy in TD

Horrobin first suggested that a deficiency of prostaglandin (PGE_1) may play an important role in the aetiology of schizophrenia (Horrobin 1979). This suggestion was based on a number of empirical observations including:

1. schizophrenics are resistant to pain and rarely suffer from rheumatoid arthritis, in which there is an increased PG synthesis
2. neuroleptics used in schizophrenia elevate prolactin levels, which in turn stimulate tissue PG synthesis
3. epilepsy and fever temporarily remit schizophrenic symptoms, and in both these conditions there is an increased PG synthesis.
4. platelets from schizophrenics fail to synthesise PGE1, when stimulated by ADP
5. clozapine, an effective antipsychotic, behaves as a PG analogue in both in vitro studies and animal behavioural tests.

Attempts have been made to increase PGE_1 synthesis to alleviate symptoms of schizophrenia by giving PG precursors, namely EFAs in the form of evening primrose oil(Vaddadi 1979). During these studies I first noted that some patients who had TD showed clinical improvement in their orofacial movements. This prompted me to see if such an effect of EFA supplementation would be seen in animal model studies of TD. We studied the effect of evening primrose oil on peri-oral movements induced by 2-(N, N-dipropyl) amino-5, 6-dihydroxy-tetralin (tetralin) in guinea pigs. The treatment with EFA abolished tongue protrusion completely but the reductions in grimacing, munching and biting behaviour were small, but significant. We concluded that EFA, by increasing brain PG production, in turn may reduce the dopaminergic function or alter GABA-ergic or serotonergic function and thus reduce tetralin-induced peri-oral movements (Nohria and Vaddadi 1982).

We observed similar reduction in dyskinesia in our dihomo-gamma-linolenic acid clinical trial in neuroleptic resistant schizophrenic patients (Vaddadi et al. 1986; Vaddadi 1984). Kaiya also noticed a reduction in dyskinesia when he treated schizophrenic patients with intravenous prostaglandin E1 (Kaiya 1984).

In a well-conducted animal model study, Costall et al. (1984) observed the antidyskinetic effect of synthetic dihomo-gamma-linolenic acid (DGLA), a PGE_1 precursor, by giving it intraperitoneally and by direct injection into the striatum and through the diet. DGLA was shown to exert a specific antidyskinetic action in both the rat and guinea pig. When dyskinesias were induced by injecting dopamine into the striatum (guinea pig), and in the guinea-pig and rat by injecting tetralin, the dyskinesias could be effectively reduced or abolished by DHLA given intraperitoneally, orally or directly into the brain area thought to control dyskinesia responding, i. e., the striatum. Even DGLA given in the diet for 10–14 days was shown to suppress the dyskinesias. DGLA did not modify other behavioural paradigms of striatal dopamine dysfunction. There was a need for repeated treatment with DHLA for its action, and this finding is consistent with its PG-precursor role. Aspirin, a cyclo-oxygenase inhibitor, was able to block the action of DGLA. The authors suggest that the DGLA, by getting incorporated into lipid membranes and getting converted to PG, exerts an antidyskinetic effect. These observations are consistent with our previous animal model studies (Nohria and Vaddadi 1982). These observations convinced us that EFAs may have a curative and preventative role in the early stages of neuroleptic-associated TD in humans.

Clinical trial using linoleic and gamma-linolenic acid (EAMOL) in TD in psychiatric patients

These open clinical and animal model studies were encouraging and we decided to do a double-blind, placebo-controlled study of EAMOL in patients with neuroleptic-induced TD (Vaddadi et al. 1989). Each EAMOL capsule contains linoleic acid, 360 mg, gamma-linolenic acid, 45 mg and vitamin E, 10 IU.

The study included 39 (81.3 %) schizophrenics, 5 cases of bipolar affective disorder and four with personality disorder, diagnosed according to DSM III criteria. There were 31 men and 17 women, mean age 52.7 years. They had been exposed to neuroleptics for more than three months and had well-established movement disorder of at least mild severity (with an AIMS score of 2 or more and Simpson TD rating score of 3 or more on any one item; Schooler and Kane 1982). Medication was kept constant. Blood samples were taken from 20 age-matched non-psychiatric hospital controls, together with 17 diagnostically matched hospitalised psychiatric controls without abnormal involuntary movements. Of these, 82.3 % suffered from schizophrenia and the other three had personality disorders.

After an initial acclimatisation period of four weeks on placebo capsules, the subjects were randomly allocated to receive either active EAMOL capsules or placebo EAMOL capsules for 16 weeks. Vitamin E, 10 IU, was added to both active and placebo capsules. Patients were given 12 capsules of EAMOL daily. At the end of a 16-week period, there was a crossover to the other treatment for a further period of 16 weeks. Movement disorder was assessed with the AIMS and Simpson TD scales (Guy 1976; Simpson et al. 1979). The Comprehensive Psychopathological Rating Scale (CPRS; Asberg et al. 1978) was administered and blood samples were collected regularly for red blood cell (RBC) membrane phospholipid analyses. AIMS and Simpson tests were administered at two weekly intervals and CPRS at eight weekly intervals. Table 1 shows that psychiatric patients had values of cis-linoleic acid (N-6 series) below those of normal controls. Patients who had severe TD (total AIMS>12) had the lowest

Table 1. Mean value of RBC membrane EFA levels in phospholipids

	n-3 Series (mean + SD)	n-6 Series (mean + SD)
Severe tardive dyskinesia (total AIMS>12) n=14, mean age = 50.3 yr	4.41 + 4.39[a*, b, c]	9.12 + 7.14[a*, b*]
Moderate tardive dyskinesia (total AIMS 10–12) n=11, mean age = 52.4 yr	7.37 + 5.02[a]	14.03 + 8.19[a*]
Mild tardive dyskinesia (total AIMS<10) n=12, mean age = 50.9 yr	8.37 + 4.48[a]	14.50 + 7.28[a*]
Psychiatric controls n=17, mean age = 43.1 yr	7.72 + 4.62[a]	16.00 + 7.28[a]
Normal controls n=20, mean age = 52.3 yr	11.17 + 1.74	22.00 + 2.42

[a] Mean significantly lower than normal control, $P<0.05$; [a*] $p<0.01$.
[b] Mean significantly lower than psychiatric control, $P<0.05$; [b*] $P<0.01$.
[c] Mean significantly lower than mild tardive dyskinesia, $P<0.05$.

values of N-3 and N-6 series when compared to psychiatric controls and normals.

Since 81.3 % of the subjects suffered from schizophrenia, the results suggest a close association among low EFA levels, TD and schizophrenia. When the individual RBC fatty acids were looked at in different groups, psychiatric hospital controls (primarily schizophrenics) without TD had levels of N-6 and N-3 fatty acids below normal. The difference was significant in the case of arachidonic acid (AA) 20:4N-6, 22:4N-6, 22:5N-3 and docosahexaenoic acid (22:6N-3). However, for the main dietary EFA – linoleic acid (18:2N-6) and alpha-linolenic acid (18:3 N-3) – the differences between normals and hospitalised psychiatric patients were trivial, indicating there was no dietary or absorption-related problem. Altered values of their metabolites suggest that their is an abnormality in EFA metabolism. However, there were no significant effects of EFA supplementation on TD ratings, although there was an improvement on the items of the schizophrenia sub-scale of CPRS. EFAMOL supplementation elevated dihomo-gamma-linolenic acid (DGLA) and AA levels in RBC membrane. At the end of the 16-week period all of the patients received active EFAMOL and active EFAVIT tablets. EFAVIT tablets contain ascorbic acid, 125 mg, pyridoxine hydrochloride, 25 mg, nicotinic acid, 7.5 mg, and zinc sulphate, 4.4 mg. These micronutrients are known co-factors in the metabolism of EFAs to PGs. The results of the final phase of the study are given in Tables 2 and 3. Addition of two Efavit tablets daily, combined with the active EAMOL capsules, produced a significant improvement in both mental state and TD scores after four weeks.

There was a significant elevation of N-6 and N-3 series fatty acids in RBC membrane in the absence of EFAMOL containing any N-3 series fatty acids (Table 3). It suggests that the addition of co-factors may be important in getting EFA into the RBC membrane. We believe similar membrane fatty acid changes may also be happening in the brain.

Table 2. Mean clinical ratings on the AIMS and Simpson Tardive Dyskinesia Scores and on the CPRS score comparing the end of the double-blind eight-months trial with the results of Efamol and Efavit treatment at the end of this period[a]

	AIMS	Simpson	CPRS
End of eight months (placebo last)	10.7 ± 5.7	42.3 ± 4.9	11.9 ± 6.4
Efamol/Efavit (placebo last)	7.3 ± 4.2	39.7 ± 4.8	8.5 ± 3.6
End of eight months (Efamol last)	9.8 ± 2.7	39.0 ± 3.6	9.3 ± 6.0
Efamol/Efavit (Efamol last)	6.2 ± 3.4	38.1 ± 4.0	8.4 ± 5.0
Significance (P value), end of end months vs. end of Efamol/Efavit	<0.02	<0.05	<0.01

[a] The results are shown separately for the patients who, in the double-blind phase, were receiving placebo or Efamol last. Values are mean ± SD. Statistically analysis was carried out using analysis of variance, comparing the results at the end of the eight-months trial in all patients with the results at the end of the Efamol + Efavit phase.

Table 3. Mean values of the red cell membrane and plasma phospholipid EFA levels during the whole trial period (mg/100 mg total lipid, n=18)

EFAs	Baseline month 1 mean (SD)	Placebo month 5/9 mean (SD)	Efamol month 5/9 month 10	Efamol + Efavit
RCB membrane				
n-6 Series	10.1 (6.8)	14.1 (4.7	14.3 (6.4)	20.1 (5.0)[a]
n-3 Series	4.9 (4.0)	6.9 (4.4)	6.0 (3.5)	10.6 (4.2)[a]
Plasma				
n-6 Series (2.4)[a]	11.5 (1.9)	13.0 (2.0)	14.9 (2.9)[a]	15.6
n-3 Series (2.4)	8.7 (2.7)	8.1 (4.6)	6.6 (2.0)	8.0

[a] Inidcates significant difference from baseline measures (P<0.05).

The relative failure in the EFAMOL-alone period in improving TD may be due to inadequate dosage, shorter trial period and chronicity of TD. However, the addition of PG synthesis co-factors to EFA supplementation suggests that EFA alone may be insufficient in their eventual effectiveness in improving TD.

In another study in a completely different geographical location (Melbourne, Australia), we again looked at the relationship between RBC membrane fatty acids and clinical ratings of TD in a younger schizophrenic population (Vaddadi et al., submitted for publication). Forty-seven schizophrenics and schizo-affective (DSM III criteria) inpatients were examined and a further 23 outpatient schizophrenics were also included. All had taken neuroleptics for more than three months. The mean age of the sample was 35.4 years. They were ill for an average of 12.3 years. Simpson Tardive Dyskinesia Ratings (STDRS) and Brief Psychiatric Rating Scale (BPRS) were administered and blood samples taken for RBC membrane fatty acid analysis.EFA data from 31 psychiatric patients without TD were compared to those with mild TD (N=12; total score below 50 but at least one dyskinetic symptom present) and those (N=20) with more severe TD (total score an STRDRS greater than or equal to 50). EFA from the 17 controls were included as a normal comparison sample.

Multivariate analysis showed that age had no significant multivariate effect on EFA levels. However, multivariate tests of significance for the effect of TD status revealed a highly significant result. (Pillais' approx. F Value 2.77, df 222, P<.001). Separate univariate analyses were then performed for each of the nine fatty acids examined which revealed significant effects of clinical status on gamma-linolenic acid (18:3N-6) and marginally significant effects on alpha-linolenic acid (18:3N-3) and arachidonic acid (20:4N-6).

Discriminant function analysis was performed to explore the relationship between fatty acid profiles and "clinical group membership" (schizophrenic with severe TD, no TD and normal control). Although three discriminant functions emerged, only the first function was statistically significant. This function placed

Table 4. Loading of EFA values on first discriminant function.

Variable	Loading on function 1
GLA (18:3N-6)	+ .713[a]
ALA (18:3N-3)	+ .310[a]
Adrenic (22:4N-6)	+ .127
LA (18:2N-6)	+ .068
EPA (20:5N-3)	− .037
AA (20:4N-6)	− .042
DHLA (20:3N-6)	− .080
DHA (22:6N-3)	− .303[a]
− 22:5N-3	− .338[a]

[a] Significant loadings

the groups along a continuum from normal, schizophrenic without TD, schizophrenic with mild TD to schizophrenic with severe TD (Table 4).

From the pattern of loadings it is clear that this function identifies as most abnormal the group with severe TD, who have low levels of gamma-linolenic acid (18:3N-6) and rather elevated levels of 22:5N-3 and 22:6N-3 EFA, and least abnormal the control group who had relatively high levels of N-6 series, particularly 18:3N-6 and lower N-3 series, particularly 22:5N-3 and 22:6N-3. Schizophrenics with motor signs of TD show low levels of N-6 series EFA in RBC membrane compared to schizophrenics without such symptoms. End metabolites of N-3 series EFAs (22:5N-3 and 22:6N-3) were increased in the most disabled group. The enzyme delta-6 desaturase has a greater affinity for N-3 series EFA (Budowski 1988). In the living organisms enzymes are constantly being synthesised and degraded. The concentration of an enzyme can be controlled by regulating either its rate of synthesis or its rate constant for degradation (Schimke 1973). It is possible that the activity of desaturases, particularly, delta-6 desaturase and delta-4 desaturase, is altered in schizophrenics who develop TD, either through their increased synthesis or lowered rate constant for degradation which may contribute to increasing the end line metabolites.

This study agrees with our previous findings (done in different geographical locations and different laboratories) in showing lower levels of some N-6 series EFAs in patients with TD compared to those patients without TD and normals.

EFAs and their possible relationship to the development and persistence of TD

There is no direct neuropathological evidence offering any clear cause for abnormal involuntary movements in schizophrenia. They appear more commonly in neuroleptic-treated patients than in the untreated population (Casey 1987). There are some questionable arguments in the findings relating to the theory of "dopamine supersensitivity" in TD. These are:

1. In animal model studies, supersensitivity to DA agonists can develop after a single injection of DA antagonist (neuroleptic)
2. Behavioral responses to DA agonist in neuroleptic-treated rodents are reversible on discontinuing neuroleptics.
3. Several of these observations in animals are not compatible with clinical TD: Symptoms without agonist provocation, individual vulnerability, late onset of TD, coexistence of TD with acute extrapyramidal syndrome and potential irreversibility of TD in humans.

There are some interesting observations in schizophrenia and TD:

1. schizophrenics with "anomalous dominance" are associated with marked underrepresentation of persistent TD (Barr et al. 1989).
2. substantial proportion of schizophrenics have a degree of ventricular enlargement (indicating brain atrophy) and/or cortical atrophy unrelated to physical treatment received or considered progressive during the illness course (Roberts 1988).
3. structures in the medial temporal lobe show greatest abnormalities, and the left hemisphere is more severely affected than the right hemisphere (Roberts 1988)
4. it has been suggested that these structural abnormalities are developmental in origin and occur in the fetus or post natally. These abnormalities may precede the onset of symptoms (Roberts 1988)
5. TD is more common in patients with negative schizophrenic symptoms (e.g., emotional withdrawal, poverty of speech). Schizophrenics with negative symptoms develop tardive dyskinesia at a younger age. It is, therefore, possible that developmental anomaly in the brain – negative features of schizophrenia particularly at a younger age, abnormal involuntary movements, dextral hand preference – may all have a common inherent "vulnerability factor" in an individual.

"Vulnerability Factor," EFA and Antioxidants

EFA are important constituents of neuronal membrane in maintaining its structure and function. Both N-3 and N-6 polyunsaturated fatty acids are essential to maintain the normal dopaminergic function in the cat caudate nucleus (Davidson et al. 1988). Concomitant administration of lithium and neuroleptics block the behavioral and biochemical dopamine hypersensitivity to DA agonist in rodents (Pert et al. 1978), Lithium at concentrations between 0.2 mM and 2 mM blocks the mobilisation of an EFA, dihomo-gamma-linolenic acid (DGLA), from the cells (Horrobin and Manku 1980). DGLA has also been shown to have antidyskinetic action in rodents and humans (Vaddadi 1984; Costall et al. 1984; Vaddadi et al. 1989). It is possible that depletion of DGLA from cells is a prerequisite for "DA supersensitivity" to develop either through the changes in the DA receptor structure and function and or secondarily through diminished PGE_1 production. It is possible that a sub-group of schizophrenics have an inherent difficulty in incorporating these EFA into neural membranes, thus affecting brain development, and this increases their vulnerability to develop earlier supersensitivity to neuroleptics and negative symptoms and TD at an early age, which is more likely to be irreversible due to cell death. The other sub-group of schizophrenics may easily incorporate EFA into the cell membrane, but have an impaired desaturase activity causing rapid depletion of EFA. These patients may be slow to develop TD and the DA supersensitivity is reversible after stopping neuroleptics. It has been shown that neuroleptic-treated patients with diabetes have an increased prevalence of TD (Ganzini et al. 1991). The activity of delta 6 desaturase is reduced with ageing and with diabetes (Horrobin 1979). Delta 6 desaturase is an important enzyme in the metabolism of EFA, and reduced activity of this enzyme may result in disturbed EFA metabolism, contributing an additional risk toward the development of TD.

EFA are highly susceptible to lipid peroxidation. Some researchers have argued for a physiological role for controlled lipid peroxidation. It is argued that in living cells there may be a tight physiological regulation of lipid peroxidation and may have a functional role (Horrobin 1990b). It is possible that an accelerated lipid peroxidation of unknown cause during an early developmental period may contribute to the structural brain changes and subsequent development of TD, at least in a sub-group of schizophrenics. Long-term controlled clinical trials with EFAs and antioxidants in patients likely to take neuroleptics is warranted.

References

Asberg M, Montgomery S, Pervis C, Schalling D, Sedvall G. (1978) A comprehensive psychopathological rating scale. Acta Psych Scand 271: 5–27

Barnes TRE, Liddle PF, Curson DA, Patel M (1989) Negative symptoms, tardive dyskinesia and depression in chronic schizophrenia. Br J. Psych. 155: 99–103

Barr E, Mukkerjee S, Degreef G, Caracci G (1989) Anomalous dominance and persistent tardive dyskinesia. Biol. Psych. 25: 826–834

Bleuler E (1950). Dementia Praecox or the group of schizophrenias (transl. M. W. Hamilton) International Universities Press: New York

Breitbart W (1986) Tardive dyskinesia associated with high-dose intravenous metoclopramide. New Engl J. Med., 315: 518

Budowski P (1988) W-3 fatty acids in health and disease. Wld Rev Nutr Diet 57: 214–274

Cadet JL, Lohr JB, Jeste DV (1986) Free radicals and tardive dyskinesia. Trends Neurosci 9: 107–108

Casey DE (1987) Tardive dyskinesia. In: Meltzer HY (ed) Psychopharmacology: The third generation of progress. Raven Press, New York, pp 1411–1419

Costall BM, Kelly E, Naylor RJ (1984) The antidyskinetic action of dihomo-gamma-linolenic acid in the rodent. Br. J. Pharmacol 83: 733–740

Crow TJ (1989) A current view of the Type II Syndrome: Age of onset, intellectual impairment and the meaning of structural changes in the brain. Br J. Psych. 155 (Suppl 7) 15–20

Cutler NR, Post RM, Rey AC, Bunney WE jr. (1981) Depression-dependent dyskinesias in two cases of manic-depressive illness. 304: 1088–89

Davidson BC, Kurstjens NP, Patton J, Cantrill RC (1988) Essential fatty acid deprivation reduced the activity of apomorphine at presynaptic dopamine receptors modulating [3H] – dopamine release from cat caudate slices. Eur J Pharmacol 149: 317–322

Diplock AT (1983) The role of vitamin E in biological membranes. In: Biology of vitamin E. Ciba Foundation Symposium 101, London, Pitman, pp 44–45

Elkashef AM, Ruskin PE, Bacher N, Barret D (1990) Vitamin E in the treatment of tardive dyskinesia. Am. J. Psych. 147: 4, 505–506

Etsuo N (1987) Interaction of ascorbate and alpha-tocopherol. Ann. NY Acad. Sci. 498: 186–199

Fahn S (1989) The endogenous toxin hypothesis of the aetiology of Parkinson's disease and a pilot trial of high-dosage antioxidants in an attempt to slow the progress of the illness. Ann. NY Acad. Sci. 570: 186–196

Farrell PM, Levine SL, Murphy MD, Adams AJ (1978) Plasma tocopherol levels and tocopherol lipid relationships in a normal population of children as compared to healthy adults. Am. J. Clin. Nutr. 31: 1720

Ganzini L, Heintz RT, Hoffman WF, Casey DE (1991) The prevalence of tardive dyskinesia in neuroleptic treated diabetics: A controlled study. Arch. Gen. Psych 48 (3): 259–263

Guy W (1976) ECDEU Assessment Manual for Psychopharmacology, Washington, D. C., United States Department of Health, Education and Welfare

Halliwell B, Gutterdige JMC (1985) Free radicals in biology and medicine. Clarendon Press, Oxford

Hassam AG, Willis AL, Denton JP et al (1980) The effect of essential fatty acid deficient diet on the levels of prostaglandins and their fatty acid precursors in the rabbit brain. Lipids 14: 78–80

Holman RT (1966) Essential fatty acid deficiency. In: Holman RT (ed) Progress in the chemistry of fats and other lipids. Pergamon Press, New York, 275–348

Hornykiewicz O (1982) Brain neurotransmitter changes in Parkinson's disease. In: Marsden DC, Fahns (eds) Movement disorders. Butterworth Scientific, London, pp 41–58

Horrobin DF (1978) Prostaglandin and muscle. In: Horrobin DF (ed) Prostaglandins physiology, pharmacology and clinical significance. Churchill Livingstone, Edinburgh pp 77–89

Horrobin DF (1979) Schizophrenia: reconciliation of the dopamine, prostaglandin and opioid concepts and the role of pineal. Lancet, 1: 529–531

Horrobin DF (1982) Prostaglandins, Essential Fatty Acids and Psychiatric disorders: A Background Review. In: Horrobin DF (ed) Clinical uses of essential fatty acids, Montreal, Eden Press, 1982

Horrobin DF (1990 a) Essential Fatty Acids, psychiatric disorders and neuropathies. In: Horrobin DF (ed) Omega-6 essential fatty acids: Pathophysiology and roles in clinical medicine. New York, Alan R Liss, pp 305–319

Horrobin DF (1990 b) Essential fatty acids, lipid peroxidation and cancer. In: Horrobin DF (ed) Omega-6 essential fatty acids: pathophysiology and roles in clinical medicine. Alan R Liss, Inc, pp. 351–377

Horrobin DF, Manku MS (1980) Possible role of prostaglandin E_1 in the affective disorders and in alcoholism. Br Med J 1: 1363–1366

Horrobin DF, Manku MS (1990) Clinical biochemistry of essential fatty acids. In: Horrobin DF (ed) Omega-6 essential fatty acids: Pathophysiology and roles in clinical medicine. New York, Alan R Liss, pp 21–53

Howitt MK, Harvey CC, Dahn CH, Scarcy MT (1972) Relationship between tocopherol and serum lipid levels for determination of nutritional adequacy. Ann. N. Y. Acad. Sci. 203: 223–236

Jeste DV, Potkin SG, Sinha S et al. (1979) Tardive Dyskinesia – reversible and persistent. Arch Gen Psych 36: 585–590

Kaiya H (1984) Prostaglandin E_1 treatment of schizophrenia. Biol Psych 19: 457–462

Kontsevoy MP, Andrusenko I, Kostrikina YE, Zvenigordskaya VYU (1991) The clinical picture and treatment of tardive dyskinesia in old age. Paper presented at the 5th. World Congress of Biological Psychiatry, Florence, June 9–14

Kraepelin E (1907) Clinical Psychiatry. Macmillan, New York

Levin M (1986) New concepts in the biology and biochemistry of ascorbic acid. New Eng J. Med. 314: 892–902

Lohr JB, Cadet JL, Lohr M, Jeste DV, Wyatt RJ (1987) Alpha-tocopherol in tardive dyskinesia. Lancet 1: 913–914

Lohr JB, Cadet JL, Lohr M, Larson L, Wasli E, Wade L, Hylton R, Vidoni C, Jeste DV, Wyatt RJ (1988 a) Vitamin E in the treatment of tardive dyskinesia: The possible involvement of free radical mechanisms. Schizophrenia Bull 14, 2, 291–296

Lohr JG, Cadet JL, Wyatt RJ, Freed WJ (1988 b) Partial reversal of iminodipropionitrile-induced hyperkinetic syndrome in rats by alpha-tocopherol (vitamin E). Neuropsychopharmacology 1: 305–9

Marsden CD, Fahn S (1987) Problems in the dyskinesias. Mov. Disord. 2, 305–312

Mead JF, Fulco AJ (1976) The unsaturated and polyunsaturated fatty acids in health and disease. CC Thomas, Springfield

Muller DPR, Lloyd JK, Wolff OH (1983). Vitamin E and neurological function. Lancet 1: 225–27

Nohria V, Vaddadi KS (1982) Tardive dyskinesia and essential fatty acids: an animal model study. In: Horrobin DF (ed) Clinical uses of essential fatty acids. Montreal, Eden Press, 199–204

Ordway RW, Singer JJ, Walsh JV jr. (1991) Direct regulation of ion channels by fatty acids. Trends Neurosci 14, 3, 96–100

Owens-Cunningham DG, Johnstone EG, Frith DC (1982): Spontaneous involuntary disorders of movement. Their prevalence, severity and distribution in chronic schizophrenics with and without treatment with neuroleptics. Arch. Gen. Psych 39: 452–461

Pall HS, Williams AC, Blake DR, Lunec J (1987) Evidence of enhanced lipid peroxidation in the cerebrospinal fluid of patients taking phenothiazines. Lancet 1: 596–599

Pert A, Rosenblatt, Sevit C, Pert C, Bunney WE (1978) Long-term treatment with lithium prevents development of dopamine receptor supersensitivity. Science 201: 171–173

Regtop H (1983) Perspective on Vitamin E. Internat Clin Nutr Rev 3, 8–21

Roberts GW (1988) Abnormalities in brain structure in schizophrenia. Curr Opin Psych 1: 83–89

Roberts GW, Crow TJ (1987) The neuropathology of schizophrenia – a progress report. Br Med. Bull No. 3, 559–615

Schimke RT (1973) Control of enzyme levels in mammalian tissues. Adv Enzymol. 37: 135–187

Schooler NR, Kane JM (1982) Research diagnosis for tardive dyskinesia. Arch Gen Psych 39: 486–487

Shriqui, Bradwjn J, Jones B, Annable L (1991) Alpha-tocopherol in the treatment of tardive dyskinesia. Paper presented at the 5th. World Congress of Biological Psychiatry, Florence, June 9–14

Simpson GM, Angus JWS (1970) A rating scale for extrapyramidal side effects. Acta Psych Scand 212: 11–19

Simpson GH, Lee JH, Zovbok B, Gardos G (1979) A rating scale for tardive dyskinesia. Psychopharmacology 64: 171–179

Vaddadi KS (1979) Penicillin and essential fatty acid supplementation in schizophrenia. Prostaglandins Med. 12: 77–80

Vaddadi KS (1984) Essential fatty acids and neuroleptic drug-associated tardive dyskinesia: preliminary clinical observations. IRCS J Med Sci 12: 678

Vaddadi KS, Gilleard CJ, Mindham RHS, Butler RA (1986) A controlled trial of prostaglandin E_1 treatment of schizophrenia. Psychopharmacology 88: 362–367

Vaddadi KS, Courtney P, Gilleard CJ, Manku MS, Horrobin DF (1989) A double blind trial of essential fatty acid supplementation in patients with tardive dyskinesia. Psych Res 27: 313–323

Wiholm BE, Mortimer O, Boethius G, Hàgstròm JE (1984) Tardive dyskinesia associated with metoclopramide. Br. Med. J. 288: 545–7

Role of Free Radicals in Alzheimer's Disease and Down's Syndrome

P.-M. Sinet and *I. Ceballos-Picot*

Summary

Down's syndrome (DS), the most frequent of congenital birth defects, results from the presence of three chromosomes 21, instead of the normal two, in all cells of affected patients. This disease is characterized by mental retardation and the constant occurrence of Alzheimer's disease (AD)-like neuropathology. Elucidation of the biological mechanisms leading to these brain impairments in DS might provide new insights into the understanding of brain pathology in normal people.

CuZn superoxide dismutase (SOD1) is one of the genes encoded by chromosome 21. As a consequence of gene dosage excess, SOD1 activity (and protein) is increased by 50 % in all DS tissues, including brain. SOD1 is ubiquitous and catalyzes the dismutation of superoxide to oxygen and hydrogen peroxide H_2O_2. The level of SOD1 protein and mRNA is particularly high in hippocampal pyramidal neurons, which are susceptible to the degenerative processes in AD. Increased SOD1 activity might result in H_2O_2 overproduction and peroxidative damage within cells. Induction of glutathione peroxidase (GPX) in DS cells supports this concept.

To test this hypothesis, cell and animal models of SOD1 overexpression were designed. In cells transfected with the human SOD1 gene, an increase in GPX activity is observed, a situation which mimics DS. In mice transgenic for SOD1, the expression pattern of the transgene in the brain is similar to that in the human and we observe an increased peroxidation. These data, like others in the literature, support the hypothesis that an imbalance in the regulation of oxygen-derived free radical production may result in brain damage and possibly contribute to the pathogenesis of AD neuropathology.

Down's Syndrome (Trisomy 21): A Model Disease for the Study of Mental Retardation and Alzheimer's Disease

Down's syndrome (DS) or trisomy 21 is the most frequent of the birth defects in the human, afflicting around 1 newborn in 1000. In most of the cases, DS results from the presence in all cells of an extra copy of whole chromosome 21. DS is characterized by morphologic abnormalities, frequent visceral malformations (heart disease in 40 % of the cases), high incidency of leukemia, and phenotypic and biological signs of accelerated aging. Dysfunctions of the central nervous system are manifested in two pathologic features, which are encountered in all trisomy 21 patients: mental retardation and Alzheimer-like neuropathology. The extent of the decrease in intellectual quotient (IQ) is variable from one patient to another and, peculiarly, intellectual level tends to deteriorate with age. In a transversal study of DS patients (Melyn and White 1973), mean IQ was 0.58 at 1.5 years, 0.45 at 6.5 years and 0.34 at 12.5 years. Longitudinal studies (M. Prieur and J. Lejeune, personal communication) provided similar results. These data suggest that the central nervous system of trisomy 21 patients, which does not show major macroscopic and ultrastructural anomalies at birth or during early infancy, may progressively be afflicted by degenerative processes that participate in IQ deterioration. Alzheimer-like neuropathology is encountered in potentially all DS patients over age 35, with senile plaques and neurofibrillary tangles distributed in the same brain territories as in Alzheimer's disease (AD). Assessment of the AD lesions in DS patients at various ages, as compared to the age-related occurrence of AD lesions in the general population, led to the conclusion that the processes leading to AD neuropathology might start in the DS brain some 50 years earlier than in "normal brain" (Rumble et al. 1989). Thus, DS can be regarded as a disease model for identifying biological mechanisms leading to mental retardation and AD-like neuropathology. We can infer from this model that an excess of one or several genes on chromosome 21 is responsible for the pathogenesis of these features.

Mental Retardation in Down's Syndrome

Chromosome 21 is the smallest of human chromosomes (Fig. 1). DNA content can be estimated to 45 million base pairs and the expected number of genes should be around 500 – 1500. So far, 21 genes and six disease loci have been mapped on chromosome 21 (Table 1).

One methodological approach to identify genes involved in the pathogenesis of mental retardation in trisomy 21 consists of first defining the region(s) of chromosome 21 susceptible to contain these genes. This approach relies on the phenotypic and molecular analysis of rare patients with partial trisomy 21. Thus, a region called DCR (Down's syndrome chromosome region; Table 1, (Rahmani et al. 1989, 1990) is suspected to contain the genes whose triplication induces many of the phenotypic features of DS, including mental retardation. This region spans 2–3 million base pairs around the 21q22.2 band on the long arm of chromosome 21. Once this region has been defined, its genetic content has to be identified. It should also be noted that the study of patients with partial trisomy 21 indicates that one or several genes located outside the DCR on chromosome 21 may also contribute to the pathogenesis of mental retardation.

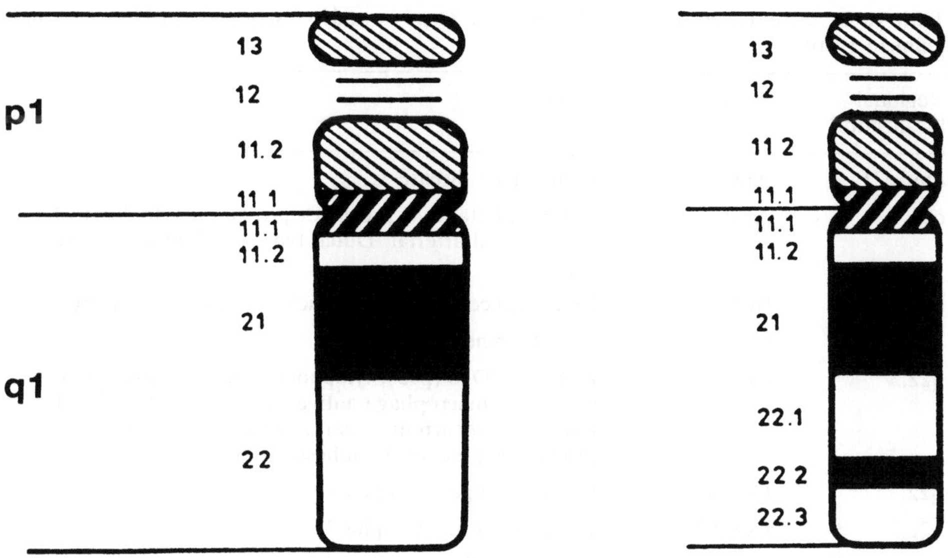

Fig. 1. Structure of chromsome 21 and banding pattern at metaphase (a) and prometaphase (b)

Alzheimer's Disease in Down's Syndrome

The AD1 Locus

Genetic linkage studies have shown that, in some families in which early-onset AD is transmitted as a dominant trait, the genetic defect lies on chromosome 21 close to the centromere (AD1 locus in Table 1, St George-Hyslop et al. 1987). It can be hypothesized that the triplication of this locus might contribute to the pathogenesis of Alzheimer-like neuropathology in DS. As the AD1 locus has not been further explored and candidate gene(s) for AD1 not identified, this hypothesis is not testable.

The Amyloid Precursor Protein Gene and Protein

The amyloid precursor protein (APP) gene has been recently cloned and assigned to chromosome 21 (Table 1; Robakis et al. 1987; Kang et al. 1987; Goldgaber et al. 1987; Tanzi et al. 1987). It encodes the protein precursor of the beta-amyloid peptide (A4) which accumulates in senile plaques and blood-vessel walls of AD and DS brains. A mutation in the nucleotidic sequence coding for A4 has been recently found in some familial AD cases, suggesting that amyloid peptide deposition is the central event in the pathogenesis of AD (Goate et al. 1991). Overexpression of the APP gene as a consequence of the presence of an APP gene extra copy might lead to the amyloid peptide

Table 1. Genes and disease loci on chromosome 21 (reviewed in Carrit and Litt 1989; Cox and Shimizu 1990)

Regional localization	Symbols	Names
qter-q21	AD1	Alzheimer's disease 1.
q21.2	APP	Amyloid beta (A4) precursor protein, amyloidosis VIb, late cerebral arterial, Dutch type (amyloid beta (A4) deposit)
q22.3	BCEI	Breast cancer, estrogen-inducible sequence expressed in
q22.3	CBS	cystathionine-beta-synthase
q22.3	CD18	Antigen CD18 (p95), lymphocyte function-associated antigen 1; macrophage antigen 1 (macl) beta subunit infections, recurrent, severe, bacterial; periodontitis, progressive; leukocyte adhesion deficiency
q22.3	COL6A1	Collagen, type VI, alpha 1
q22.3	COL6A2	Collagen, type VI, alpha 2
q22.3	CRYA1	Crystallin, alpha polypeptide 1
q22.2	DCR	Down's phenotype, with trisomy of the region
q22.3	ERG	Avian erythroblastosis virus E26 (v-ets) oncogene related
q22.3	ETS2	Avian erythroblastosis virus E26 (v-ets) oncogene homolog 2
q22.3	HMG14	High mobility group protein 14
q22.1–q22.2	IFNAR	Interferon alpha receptor
q22.1–q22.2	IFNBR	Interferon beta receptor
q22.1–q22.3	IFNGT1	Interferon gamma transducer 1
q22.3	MX1	Myxovirus (influenza) resistance 1, homolog of murine interferon-inducible protein (p78)
q22.3	MX2	Myxovirus (influenza) resistance 2, homolog of murine
q22.1	PAIS	Phosphoribosylaminoimidazole synthetase
q22.3	PFKL	Phosphofructokinase, liver type, hemolytic anemia; PFKL deficiency
q22.1	PGFT	Phosphoribosylglycinamide formyltransferase
q22.1	PRGS	Phosphoribosylglycinamide synthetase
q22.3	S100B	S100 protein, beta polypeptide (neural)
q22.1	SOD1	Superoxide dismutase 1, soluble

deposition in DS brain. Nevertheless, the processing of APP to amyloid peptide requires two proteolytic cleavages, one in the extracellular domain of the protein and the other in the membrane domain (Kang et al. 1987). In vitro experiments suggest that the physiological processing of APP does not involve these two proteolytic sites (Sisodia et al. 1990; Esch et al. 1990).

Thus, the cleavage of APP to amyloid peptide requires the intramembrane proteolytic site to be accessible to protease, which may arise as a consequence of membrane damages (Dyrks et al. 1988). This hypothesis is attractive with regard to Alzheimer-like neuropathology in DS since membrane damage could be one of the biological consequences of the rapid aging observed in this disease. Furthermore, there is now increasing evidence that free radical metabolism is disturbed in trisomy 21 as a consequence of superoxide dismutase gene dosage excess.

CuZn Superoxide Dismutase Excess and Consequences

The gene encoding CuZn superoxide dismutase (SOD1) is located on chromosome 21 near the interface between 21q21 and 21q22.1 bands (Sinet et al. 1976). SOD1 gene is expressed in all tissues but its level of expression is particularly high in neurons. Immunohistochemistry (Delacourte et al. 1988) and cDNA mRNA in situ hybridization (Ceballos-Picot et al. 1989, 1991c) studies on human hippocampus have shown that the amounts of SOD1 protein and mRNAs are high in large pyramidal neurons and granular cells, suggesting that the production of superoxide radicals within these cells might be particularly elevated. Furthermore, an analysis of Alzheimer brains indicates that the neurons in which SOD1 is high are those that are susceptible to the Alzheimer degeneration (Delacourte et al. 1988).

In trisomy 21, SOD1 activity and protein have been found to be increased by around 50% in all tissues (review by Sinet 1982), including brain (Brooksbank and Balazs 1984), as a consequence of the presence of an extra copy of the SOD1 gene. Since SOD1 catalyzes the dismutation of superoxide radicals O_2- to oxygen and hydrogen peroxide (H_2O_2), it has been hypothesized that increased SOD1 in trisomy 21 might induce overproduction of H_2O_2 within cells and accelerate oxidative damages to cell components (Sinet 1982). Supporting this hypothesis is the finding that seleno-dependent glutathione peroxidase (GPX) activity which catalyzes the reduction of H_2O_2 has been found to be increased in trisomy 21 tissues like red cells and platelets (Sinet et al. 1975; Crosti et al. 1989). In fibroblasts, normal (Feaster et al. 1977; Crosti et al. 1989) or increased (Sinet et al. 1979; Anneren and Epstein 1987). GPX activities have been reported whereas in fetal brain GPX was found to be normal (Brooksbank and Balazs 1984).

The increase in GPX activity in trisomy 21 tissues does not result from a gene dosage effect, since the gene encoding GPX is on chromosome 3. This GPX increase is actually linked to SOD1 excess. As a matter of fact, transfection of the human SOD1 (hSOD1) gene into human and mouse cells in culture induces an increase in GPX activity (Ceballos-Picot et al. 1988; Kellner and Bagnell, 1990). It seems likely that the induction of GPX resulting from SOD1 excess is mediated via an overproduction of H_2O_2 within cells. Other experiments support this interpretation. Human and mouse cells transfected with hSOD1 showed enhanced lipid peroxidation (Elroy-Stein et al. 1986; Elroy-Stein and Groner 1988) and hSOD1 rat PC12 cells have impaired neurotransmitter uptake resulting from membrane modifications of the chromaffin granules which may result from lipid peroxidation (Elroy-Stein and Groner 1988).

To further explore the deleterious effects of excess SOD1, transgenic mice for hSOD1 were obtained by us (Ceballos et al. 1991 a, b) and other group (Epstein et al. 1987). In our transgenic strain, one hSOD1 copy had been integrated in the mouse genome and SOD1 activity was higher than in control mice in almost all tissus (Ceballos-Picot et al. 1991 a, b). In brain, the SOD1 activity is increased by a factor of 2 and the neurone specificity of the hSOD1 expression is similar to that observed for human SOD1, i.e. in the hippocampus large pyramidal neurons and granular cells (Ceballos-Picot 1991 a). We also observed in transgenic brains a significantly higher basal amount of thiobarbituric acid material, strongly suggesting an increased level of peroxidation in vivo (Ceballos-Picot et al. 1991 a). Another interesting feature that has been observed in hSOD1 transgenic mice is ultrastructural changes in the neuromuscular junction of the tongue muscle similar to those seen in the muscle of aging mice and rats, as well as in tongue muscles of DS patients (Avraham et al. 1988). Whether or not these anomalies could be linked to increased peroxidative damage remains to be proven. Nevertheless, this observation suggests that SOD1 excess could have deleterious effects on the nervous system. Extended biochemical and ultrastructural studies of the brain of hSOD1 transgenic mice should determine whether SOD1 excess participates in the accelerated aging processes in DS, expecially with regard to Alzheimer-like neuropathology. Moreover, study of this animal model has general implications, particularly with respect to the potential role of free radicals in aging and age-related diseases of the central nervous system.

References

Anneren G, Epstein CJ (1987) Lipid peroxidation and superoxide dismutase-1 and glutathione peroxidase activities in trisomy 16 fetal mice and human trisomy 21 fibroblasts. Pediatr Res 21: 88–92

Avraham KB, Schickler M, Sapoznikow D, Yarom R, Groner Y (1988) Down's syndrome: abnormal neuromuscular junction in tongue of transgenic mice with elevated levels of human CuZn-superoxide dismutase. Cell 54: 823–829

Brooksbank BWL, Balazs R (1984) Superoxide dismutase, glutathione peroxidase and lipoperoxidation in Down syndrome fetal brain. Dev Brain Res 16: 37–44

Carrit B, Litt M (1989) Report of the committee on the genetic constitution of chromosomes 20 and 21. Cytogenet Cell Exenet 51: 358–371

Ceballos-Picot I, Delabar JM, Nicole A, Lynch RE, Hallewell RA, Kamoun P, Sinet PM (1988) Expression of transfected human CuZn superoxide dismutase gene in mouse L cells and NS20Y neuroblastoma cells induces enhancement of glutathione peroxidase activity. Biochim. Biophys. Acta 949: 58–64

Ceballos-Picot I, Javoy-Agid F, Hirsch EC, Dumas S, Kamoun P, Sinet PM, Agid Y (1989) Localisation of copper-zinc superoxide dismutase mRNA in human hippocampus by in situ hybridization. Neurosci. Lett. 105: 41–46

Ceballos-Picot I, Nicole A, Briand P, Grimber G, Delacourte A, Defossez A, Javoy-Agid F, Lafon M, Blouin JL, Sinet PM (1991 a) Neuronal specific expression of human copper-zinc superoxide dismutase in transgenic mice: animal model of gene dosage effect in Down's syndrome. Brain Res. 552: 198–214

Ceballos-Picot I, Nicole A, Briand P, Grimber G, Delacourte A, Flament S, Blouin JL, Thevenin M, Kamoun P, Sinet PM (1991 b) Expression of human Cu-Zn superoxide dismutase gene in transgenic mice: model for gene dosage effect in Down syndrome. Free Radic Res Commun 12–13: 581–589

Ceballos-Picot I, Javoy-Agid F, Delacourte A, Defossez A, Lafon M, Hirsch E, Nicole A, Sinet PM, Agid Y (1991 c) Neuronal localization of copper-zinc superoxide dismutase protein and mRNA within the human hippocampus from control and Alzheimer's disease brains. Free Radic Res Commun 12–13: 571–580

Cox DR, Shimizu N (1990) Report of the committee on the genetic constitution of chromsome 21. Cytogenet Cell Genet 55: 235–244

Crosti N, Bajer J, Gentile M, Resta G, Serra A (1989) Catalase and glutathione peroxidase activity in cells with trisomy 21. Clin. Genet. 36: 107–116

Delacourte A, Defossez A, Ceballos I, Nicole A, Sinet PM (1988) Preferential expression of copper zinc superoxide dismutase in the vulnerable cortical neurons in Alzheimer's disease. Neurosci. Lett. 92: 247–253

Dyrks T, Weidemann A, Multhaup G, Salbaum JM, Lemaire HG, Kang J, Muller-Hill B, Masters CL, Beyreuther K (1988) Identification, transmembrane orientation and biogenesis of the amyloid A4 precursor of Alzheimer's disease. EMBO J. 7/4: 949–957

Elroy-Stein O, Groner Y (1988) Impaired neurotransmitter uptake in PC12 cells overexpressing human Cu-Zn superoxide dismutase: implications for gene dosage effects in Down's syndrome. Cell 52: 259–267

Elroy-Stein O, Bernstein Y, Groner Y (1986) Overproduction of human Cu-Zn superoxide dismutase in transfected cells: extenuation of paraquat-mediated cytotoxicity and enhancement of lipid peroxidation. EMBO J. 5: 615–622

Epstein CJ, Avraham KB, Lovett M, Smith S, Elroy-Stein O, Rotman G, Bry C, Groner Y (1987) Transgenic mice with increased CuZn superoxide dismutase activity: animal model of dosage effects in Down syndrome. Proc. Natl. Acad. Sci. USA 84: 8044–8048

Esch FS, Keim PS, Beattie EC, Blacher RW, Culwell AR, Oltersdorf, McClure D, Ward PJ (1990) Cleavage of amyloid beta peptide during constitutive processing of its precursor. Science 248: 1122–1124

Feaster WW, Kwok LW, Epstein CJ (1977) Dosage effects for superoxide dismutase-1 in nucleated cells aneuploid for chromosome 21. Am. J. Hum. Genet. 29: 563–570

Goate A, Chartier-Harlin MC, Mullan M et al. (1991) Segregation of a missense mutation in the amyloid precursor protein gene with familial Alzheimer's disease. Nature 349: 704–706

Goldgaber D, Lerman MI, McBride W, Saffioti V, Gajdusek DC (1987) Characterization and chromosomal localization of a cDNA encoding brain amyloid of Alzheimer's disease. Science 235: 877–880

Kang J, Lemaire HG, Unterbeck A, Salbaum JM, Masters C, Grezschik KH, Multhaup G, Beyreuther K, Muller-Hill B (1987) The precursor of Alzheimer's disease amyloid A4 protein resembles a cell-surface receptor. Nature 325: 733–736

Kellner MJ, Bagnell R (1990) Alteration of endogenous glutathione peroxidase. Manganese superoxide dismutase and glutathione peroxidase activity in cells transfected with a copper-zinc superoxide dismutase expression vector. J. Biol. Chem. 265/19: 10872–10875

Melyn MA, White DT (1973) Mental and developmental milestones of noninstitutionalzed Down's syndrome children. Pediatrics 52: 542–545

Rahmani Z, Blouin JL, Créau-Goldberg N, Watkins PC, Mattei JF, Poissonnier M, Prieur M, Chettouh Z, Nicole A, Aurias A, Sinet PM, Delabar JM (1989) Critical role of the D21S55 region on chromosome 21 in the pathogenesis of Down syndrome. Proc Natl Acad Sci USA 86: 5958–5962

Rahmani Z, Blouin JL, Créau-Goldberg N, Watkins PC, Mattei JF, Poissonnier M, Prieur M, Chettouh Z, Nicole A, Aurias A., Sinet PM, Delabar JM (1990) Down syndrome critical region around D21S55 on proximal 21q22.3. Am J Med Genet 7 [Suppl]: 98–103

Robakis NK, Ramakrishna N, Wolfe G, Wisniewski HM (1987) Molecular cloning and characterization of a cDNA encoding the cerebrovascular and the neuritic plaque amyloid peptides. Proc. Natl. Acad. Sci. USA 84: 4190–4194

Rumble B, Retallack R, Hilbich C et al. (1989) Amyloid A4 protein and its precursor in Down's syndrome and Alzheimer' Disease. N Engl J. Med. 22: 1446–1452

Sinet PM (1982) Metabolism of oxygen derivatives in Down's syndrome Ann. N.Y. Acad. Sci. 386: 82–94

Sinet PM, Michelson AM, Bazin A, Lejeune J, Jérome H (1975) Increase in glutathione peroxidase activity in erythrocytes from trisomy 21 subjects. Biochem. Biophys. Res. Commun 67: 910–915

Sinet PM, Couturier J, Dutrillaux et al. (1976) Trisomie 21 et superoxide dismutase-1 (IPO-A). Tentative de localisation sur la sous-bande 21q22.1. Exp Cell Res. 97: 47–55

Sinet PM, Lejeune J, Jérome H (1979) Trisomy 21 (Down's syndrome), glutathione peroxidase, hexose monophosphate shunt and I. Q. Life Sci 24: 29–34

Sisodia SS, Koo EH, Beyreuther K, Unterbeck A, Price DL (1990) Evidence that beta-amyloid protein in Alzheimer's disease is not derived by normal processing. Science 248: 492–495

St George-Hyslop PH, Tanzi RE, Polinsky RJ et al. (1987) The genetic defect causing familial Alzheimer's disease maps on chromosome 21. Science 235: 885–890

Tanzi RE, Gusella JF, Watkins PC, Bruns GA, St George-Hyslop P, Van Keuren ML, Patterson SP, Kurnit DM, Neve RL (1987) Amyloid beta protein gene: cDNA, mRNA distribution and genetic linkage near the Alzheimer locus. Science 235: 880–884

Dopaminergic Cell Death in Parkinson's Disease

F. Javoy-Agid

Introduction

The main biochemical characteristic of Parkinson's disease is the degeneration of the dopaminergic nigrostriatal neurons which originate in the substantia nigra pars compacta and project to the striatum (caudate nucleus and putamen; Agid et al. 1987). The cause of the neuronal death remains unknown. The origin of the disease does not seem to be due to some viral infection or autoimmune phenomena. Studies on monozygotic twins have found a low concordance rate, suggesting a limited role for heredity (Duvoisin 1986). However, there might be a genetic component, since an autosomal dominant transmission has been described in some families (Johnson et al. 1990). An endogenous factor may contribute to the appearance of the disease; indeed, abnormalities in the mitochondrial electron transport chain have been observed in idiopathic Parkinson's disease, suggesting that the mitochondrial genome might be affected (Parker et al. 1989; Schapira et al. 1990a). However, one cannot exclude the possibility that an exogenous factor is responsible for the disease, and epidemiological investigations have searched for environmental risk factors (Koller et al. 1990). The hypothesis for an exogenous toxis factor has been supported by observations of cases of permanent parkinsonism produced in humans and monkeys by 1-methyl-4-phenyl-1,2,3,6-tetrahydropyridine (MPTP) intoxication, a drug thought to selectively destroy the nigrostriatal dopaminergic neurons (Langston et al. 1983). Therefore, dopaminergic cell death in Parkinson's disease could be due to the presence of a permanent exogenous or endogenous toxic factor. The biochemical characteristics of the striatal and nigral dopamine denervation, as well as a number of chemical abnormalities detectable at the time of decease in the substantia nigra, support this hypothesis.

Characteristics of the Dopaminergic Denervation in Striatum of Patients with Parkinson's Disease

The diagnosis of Parkinson's disease is classically confirmed by neuropathological evidence of cell loss in the substantia nigra. The loss of these neurons leads to a dramatic decrease in striatal dopamine and tyrosine hydroxylase levels (Agid et al. 1987). A gradient in the deficit of dopamine transmission relative to that in control brains has been demonstrated by both biochemical (Kish et al. 1988; Schermann et al. 1989) and immunocytochemical (Graybiel et al. 1990) methods: the putamen is more severely affected than the caudate nucleus, which is more affected than the ventral striatum. In addition, dopamine-containing denervation is greater in the more dorsal parts of the striatum than in the more ventral parts. The data indicate that Parkinson's disease affects the dopaminergic neuronal cells differently.

The characteristics of the striatal dopaminergic degeneration as a function of the duration of the disease were investigated using as a specific marker, the binding of 3H-dihydrotetrabenazine (3H-TBZOH), a ligand of vesicular monoamine transporter (Schermann et al. 1989). This study showed that 3H-TBZOH binding was always subnormal in parkinsonian brains postmortem (Fig. 1). However, binding levels could be more affected in some patients with a short duration of disease than in other patients with a longer duration of the disease, indicating that the rate of nerve cell death is variable from patient to patient. In addition, the levels of 3H-TBZOH binding decreased through the course of the disease, suggesting a continuous progression of the dopaminergic denervation. Compared to the rate of dopamine denervation as a function of age in the control population, the decrease progressed about twice as rapidly in Parkinson's

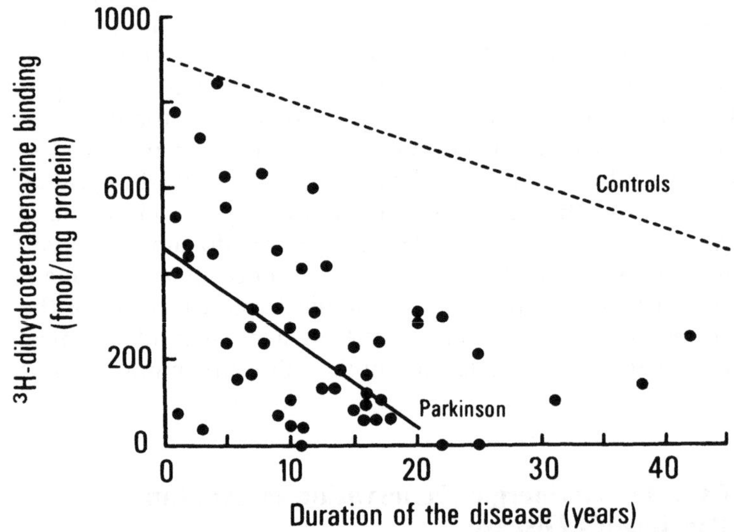

Fig. 1. Levels of 3H-dihydrotetrabenazine binding in caudate nucleus of parkinsonian patients as a function of duration of the disease. Striatal dopaminergic innervation was investigated postmortem using a specific marker, the binding of 3H-dihydrotetrabenazine, which is a ligand of vesicular monoamine transporter. The rate of dopamine denervation in Parkinson's disease was compared to the rate of denervation as a function of age in the control (dotted line) population. Black dots represent individual values for parkinsonian patients (representation of data from Schermann et al. 1989)

disease than in normal aging. Aging does not seem to play a major role in the process of degeneration of the nigrostriatal neurons in Parkinson's disease: the rate of denervation was similar in patients with an early age of onset of the disease (before the age of 60) as in those whose disease started later on (after the age of 60) (Schermann et al. 1989).

Thus, in Parkinson's disease, the progression of dopaminergic cell death in the substantia nigra is slow but more rapid than in normal aging. This indicates that the cause of degeneration is continuously present, or that a vital process has been once and for all disrupted, perhaps at the level of the genome, causing insufficient production of essential proteins or accumulation of eventually toxic metabolites.

The finding that dopamine denervation may be due to a permanent insult is supported by the presence of abnormalities seen at the time of death at the level of the parkinsionian substantia nigra.

Neurochemical Abnormalities in the Substantia Nigra in Parkinson's Disease

Identification of dopaminergic neurons by tyrosine hydroxylase immunocytochemistry clearly demonstrates that the cell loss is variable from patient to patient (Hirsch et al. 1991a). but never reaches the total population of these neurons (Hirsch et al. 1988). The remaining dopaminergic neurons are neurons which may not be intact, but are most likely suffering. 1) A type of dopaminergic neuron, never observed in control substantia nigra, has been identified in Parkinson's disease, characterized by the presence of neuromelanin pigment (a reliable marker for dopaminergic neurons in the human substantia nigra; Hirsch et al. 1988) in the absence of detectable tyrosine hydroxylase immunoreactivity. These pigmented tyrosine hydroxylase-negative cells may represent dying neurons that once contained the enzyme. 2) Surviving dopaminergic neurons express subnormal levels of messenger RNA (mRNA) for tyrosine hydroxylase, as evidenced by in situ hybridization using a radioactively labelled probe (Javoy-Agid et al. 1990), which may result from changes in either tyrosine hydroxylase gene transcription or modification at the post-transcriptional level. The change may be a consequence of a disruption of transneuronal regulation of the activity of nigral neurons. A number of neuronal systems that are considered to influence the activity of nigral dopaminergic neurons are affected in Parkinson's disease (Agid et al. 1987). Alternatively, the decreased amount of nigral tyrosine hydroxylase mRNA may simply reveal the suffering of the dopaminergic neurons still present at late stages of the disease. 3) Six times as many HLA-DR (a human glycoprotein) -positive microglia, phagocytosing dopaminergic neurons, have been counted in the parkinsonian substantia nigra compared to controls, indicating that a still-active pathological process is affecting dopaminergic cells at the end stages of the pathology (McGeer et al. 1988).

The hypothesis that dopamine denervation may be due to a permanently ongoing insult is supported by the increase in basal level of lipid peroxidation in the parkinsonian substantia nigra postmortem. This increase is suggested both by the enhanced levels in malondialdehyde (an intermediate in the lipid peroxydation process) content and the reduction in polyunsaturated fatty acid levels (an index of the amount of substrate for lipid peroxidation; Dexter et al. 1986, 1989a, 1990). These biochemical changes, evidenced specifically in the substantia nigra (they were not observed in other brain areas such as caudate nucleus, putamen, pallidum), may indicate that an increased rate of lipid peroxidation continues to occur up to the time of death, perhaps due to continued exposure to excess free radicals derived from some endogenous or exogenous neurotoxic species.

A number of factors may possibly be related to dopaminergic cell death in Parkinson's disease, including the following.

Presence of neuromelanin in dopaminergic neuron

Several dopamine-containing cell groups in the human mesencephalon have been shown using tyrosine hydroxylase immunocytochemistry. Tyrosine hydroxylase-positive neurons densely distribute ventrally within the substantia nigra pars compacta and are present laterally in the substantia nigra pars lateralis, the peri- and retrorubral tegmental cell group A8 and ventral tegmental area (A10), the paramedium tegmentum and the central gray substance (Hirsch et al. 1988). Differences among the characteristics of these cell groups are suspected , not only because of their distribution and their connections (Graybiel et al. 1990), but also because of some of their biochemical characteristics. For example, two types of neurons have been identified depending on whether they do or do not contain neuromelanin. The proportion of melanized neurons in different dopaminergic cell groups within the normal human midbrain varies from very high (for neurons in the substantia nigra pars compacta) to intermediate levels for the ventral tegmental region and the peri- and retrorubral region, to very low levels for the tyrosine hydroxylase-containing cells in the central gray substance (Hirsch et al. 1988; Fig. 2). In Parkinson's disease, quantitative analysis has shown that, among total populations of catecholaminergic neurons in the midbrain, subpopulations of pigmented neurons were the most vulnerable (Fig. 3). Indeed, the higher the percentage of melanized neurons in the dopaminergic cell group, the greater the loss of neurons in that cell group, as judged by comparisons with controls (Hirsch et al. 1988, 1989). Thus Parkinson's disease does not affect all dopaminergic neurons, but it preferentially affects the neuromelanized neurons. These findings suggest that neuromelanin may be related directly or indirectly to the pathological process underlying cell death in this pathology. Mechanisms proposed as links between neuromelanin content and vulnerability of dopaminergic neurons include the generation of free radicals in the synthesis of neuromelanin (Marsden 1983; Graham 1979; Graham et al. 1978) and the binding by neuromelanin of neurotoxins such as MPP^+, the toxic metabolite of MPTP. Accumulation of neuromelanin could be gradual and the production of free radicals or release of toxic agents could be paced proportionally.

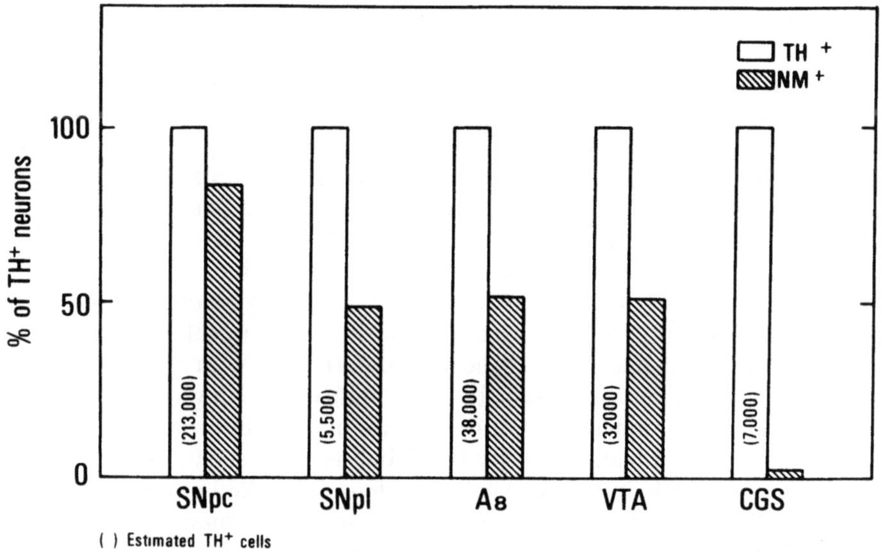

Fig. 2. Percentage of melanized dopaminergic neurons in the different doaminergic cell groups of the human mesencephalon. Dopaminergic neurons (TH^+) were identified by immunocytochemistry using a tyrosine hydroxylase antibody. In each dopaminergic cell group, the number of dopaminergic neurons containing neuromelanin (NM^+) is expressed as a percentage of total number of dopaminergic cells (= 100 %) in each area, respectively. In brackets, absolute number of neurons. Substantia nigra pars compacta (SNpc), pars lateralis (SNpl); peri- and retrorubral tegmentum (A8); ventral tegmental area (VTA); central gray substance (CGS) (schematic representation of original data from Hirsch et al. 1988)

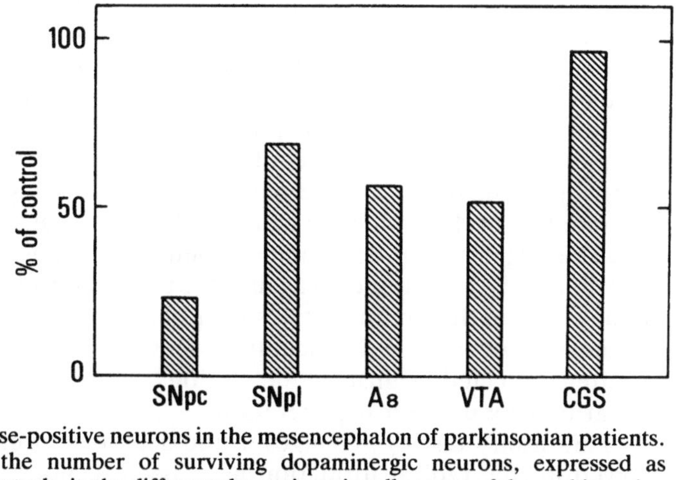

Fig. 3. Tyrosine hydroxylase-positive neurons in the mesencephalon of parkinsonian patients. Quantitative analysis of the number of surviving dopaminergic neurons, expressed as percentage of respective controls, in the different dopaminergic cell groups of the parkinsonian mesencephalon. For abbreviations, see legend to Figure 2 (Data are the schematic representation of absolute values reported by Hirsch et al. 1988)

Protective enzymes against oxygen toxicity

Oxygen free radicals have been implicated as a potential cytotoxic mechanism responsible for nigral cell death in Parkinson's disease (Calne and Langston 1983). Several enzymes act in concert in defense against oxygen toxicity. Dismutation of superoxide radical to hydrogen peroxide is catalyzed by superoxide dismutase (SOD). The resulting hydrogen peroxide is then decomposed to water by catalase or removed by glutathione peroxidase (GPX), which uses hydrogen peroxide to oxidize the reduced glutathione. The neuromelanin-pigmented neurons in the control human mesencephalon may have specific characteristics in this defense system, and the biochemical pathways leading to oxygen species formation may be particularly active in these cells. Indeed, neuromelanized neurons in the substantia nigra express high levels of mRNA for CuZnSOD, as evidenced by autoradiographic detection at the cellular level using in situ hybridization methods (Ceballos et al. 1990). Within the human midbrain, positively hybridized cells are densely packed in the substantia nigra pars compacta. Some are also present in the ventral tegmental area and, in these areas, they correspond greatly to neuromelanin-containing neurons (Ceballos et al. 1990). Thus the CuZnSOD gene is preferentially and highly expressed in the neurons that are most vulnerable to the pathological process in Parkinson's disease. The neuromelanin neurons may need a high CuZnSOD content to facilitate removal of superoxide radicals. Alternatively, by promoting hydrogen peroxide production, a high cellular CuZnSOD activity might contribute to the vulnerability of these neurons. The preferential localization of CuZnSOD mRNA in neurons is all the more interesting if one considers the cellular localization of GPX, the enzyme-scavenging hydrogen peroxide generated by cellular metabolism. Immunocytochemical studies on human mesencephalon showed the protein to be exclusively in astrocytic glial cells (Damier et al., submitted for publication). Therefore, it seems the oxygen defense system of the dopaminergic nigral neurons may require a neuronal-glial coupling. In control brains, the density of GPX-positive cells differed significantly between the four dopaminergic cell groups described above: it was high in the dopaminergic areas spared in Parkinson's disease, such as central gray substance, and low in a region characterized by a dramatic dopaminergic neuronal loss, such as the substantia nigra pars compacta. The density of GPX-positive cells was intermediate in the ventral tegmental region and the peri- and retrorubral region, characterized by a moderate loss of dopaminergic neurons (Damier et al., submitted for publication). This suggests that dopaminergic neurons surrounded by a low density of glial cells may be less protected against the production of hydrogen peroxide and this may explain their preferential susceptibility to oxidative stress. Conversely, dopaminergic neurons in the vicinity of densely concentrated glial cells seem to be better protected against the deleterious action of free radicals, which may prevent or slow up their degeneration.

In Parkinson's disease, changes in the activities of enzymes and substrates involved in free radical formation have been demonstrated on tissue homogenates of substantia nigra: an increase in SOD-like activity (Marttila et al. 1988; Saggu et al. 1989), a marked reduction in glutathione levels (Perry and Yong

1986) and of GPX activity (Ambani et al. 1975) with no change in glutathione transferase activity (Perry and Yong 1986), as well as a reduction in catalase activity (Ambani et al. 1975). The data suggest that oxygen toxicity protective mechanisms may be impaired in the substantia nigra of Parkinson's patients, and that oxygen toxicity-related events may participate in neuronal degeneration associated with the disease.

Physiological metals

The role of physiological metals has been suspected in the pathophysiology of several neurodegenerative diseases: copper for Wilson's disease (Scheinber and Sternlieb 1984), aluminium in Alzheimer's disease (Birchall and Cappell 1988), and manganese for some parkinsonian syndromes (Barbeau 1984). In Parkinson's disease, iron could participate in the mechanism underlying increased lipid peroxidation, since it produces free radicals by itself (Halliwell and Gutteridge 1986), and is selectively bound to neuromelanin to produce $Fe2+$/melanin complex, which is suspected of inducing oxidative stress (Youdim and Ben-Shachar 1989). Therefore, excessive accumulation of iron in the brain in Parkinson's disease might accelerate cell death. In tissue homogenates of parkinsonian brain, the total iron content was increased in substantia nigra, although not in any other brain region examined (caudate nucleus, putamen, cerebral cortex; Dexter et al. 1989b). These data agree with those obtained on formalin-fixed material (Earle 1968), but are discordant with those reported by Riederer et al. (1988), who showed no difference in the amount of iron in the parkinsonian brain. An increase in nigral iron content could reflect alterations of the transport of iron across the brain barrier. Studies on cerebral spinal fluid do not support this hypothesis (Pall et al. 1987). The higher-than-normal metal content could be the consequence of an increased transport of iron into nigral cells as a transferrin-iron complex via specific cell surface receptors (Bomford an Munro 1985). One cannot exclude the possibility that the accumulation of iron in substantia nigra may be a non-specific product of cellular degeneration rather than a cause. However, the absence of increase in iron content in the nigral tissue from patients with progressive supranuclear palsy, a parkinsonian syndrome also associated with a loss of dopaminergic nigro-striatal neurons (Hirsch et al. 1991b), suggests that the accumulation of iron abserved in parkinsonian substantia nigra is not merely a non-specific marker of cell death. Supporting this assumption is the comparison of the nigral tissue content in controls and parkinsonian patients by X-ray microanalysis, which showed that iron levels were increased in regions devoid of neuromelanin but decreased in neuromelanin aggregates (Hirsch et al. 1991b). Classically, iron is observed in astrocytes in the substantia nigra (Hallgren and Sourander 1985); a gliosis could therefore explain the increased iron content in the substantia nigra of patients. Iron content was decreased on neuromelanin aggregates. The fact that the surviving dopamine neurons in Parkinson's disease contain low amounts of iron bound to neuromelanin may indicate that these cells were neurons that had low amounts of iron before the disease started, which would explain their lesser vulnerability. Alternatively, the subnormal iron decrease could be a consequence of the pathological process affecting the remaining cells in the structure.

Not all forms of iron are capable of stimulating toxic radical formation. For this purpose, iron must exist in free from or as part of the intracellular mobile pool (Halliwell and Gutteridge 1986). In brain most iron is inactivated by association with ferritin (Hallgren and Sourander 1958), with relatively little in a free and reactive form. Surprisingly, considering that normally the cellular availability of iron controls ferritin synthesis (Hentze et al. 1987; Casey et al. 1988), ferritin levels in substantia nigra were markedly decreased in parkinsonian patients compared to controls (Dexter et al. 1990). Consequently, the excess iron might be present either in a free and reactive from or bound to another iron-binding molecule. It is noteworthy that ferritin levels were also decreased in many other areas of the parkinsonian brain (Dexter et al. 1990). Thus, further studies are required to establish why a reduction in brain ferritin occurs, and whether the changes in brain iron and ferritin levels observed in Parkinson's disease are related to the cause of the disease.

Conclusion

A great deal of evidence obtained postmortem on autopsy material suggests that Parkinson's disease may be the reult of a toxic process. Insight into a possible pathogenic cause of the disease was provided by recent progress in studies on the mechanisms of cell death induced by MPTP in animals. Inhibiton of complex I and of state 3 respiration by MPP^+ (the oxidation product of MPTP responsible for the neuronal toxicity) points to the mitochondria being the most probable site of action for MPP^+ (Nicklas et al. 1985). This has raised the possibility that similar abnormalities may be involved in the pathogenesis of Parkinson's disease. Data are suggestive of the of presence impaired function of the electron transport chain in parkinsonian patients. The defect seems to center around complex I (NADH-ubiquinone reductase) in the substantia nigra, but spares the striatum, suggesting that the biochemical deficiency is related to dopaminergic neuronal cells rather than their projections. This biochemical abnormality may be due to some environmental toxin or it could be the result of some endogenous toxic process in the substantia nigra, perhaps related to dopamine or neuromelanin synthesis. Complex I comprises polypeptides, some of which are encoded by mitochondrial DNA. A toxin could inhibit Complex I activity by binding to polypeptides. On the other hand, the local production of free radicals from such a toxin within the substantia nigra could account for the anatomical selectivity of the decreased Complex I activity. Since mitochondrial DNA is particularly susceptible to the mutagenic effects of free radicals (Richter et al. 1988), a defect induced in mitochondrial DNA genes encoding Complex I polypeptides could result in deficient Complex I activity. At present, analysis of mitochondrial DNA in Parkinsonian patients has failed to show any deletions in nigral tissue (Shapira et al. 1990b). In addition, the complex I defect is present in apparently unaffected tissue in patients (Parker et al. 1989). Thus, the relationship between mitochondrial abnormalities and the cause underlying dopaminergic cell death in Parkinson's disease remains to be further investigated.

References

Agid Y, Javoy-Agid F, Ruberg M (1987) Biochemistry of neurotransmitters in Parinson's disease. In: Marsden CD, Fahn S (eds.). Movement Disorders, vol 2. Butterworth, pp 166-230

Barbeau A (1984) Manganese and extrapyramidal disordes (A critical review and tribute to Dr. Georges C. Cotzias). Neurotoxicology 5:13-36

Birchall JD, Cappell JS (1988) Aluminium, chemical physiology and Alzheimer's disease. Lancet ii:1008-1010

Bomford AB, Munro HN (1985) Transferrin and its receptors: their roles in cell function. Hepathology 5:870-875

Calne DB, Langston JW (1983) Aetiology of Parkinson's disease. Lancet ii:1457-1459

Casey JL, Hentze MW, Koeller DM, Caughman SW, Rouault TA, Klausner RD, Harford J (1988) Iron responsive elements: regulatory RNA sequences that control mRNA levels and translation. Science 240:924-928

Ceballos I, Lafon M, Javoy-Agid F, Hirsch E, Nicole A, Sinet PM, Agid Y (1990) Superoxide dismutase and Parkinson's disease. Lancet 335, 8696, i:1025-1036

Dexter D, Carter C, Javoy-Agid F, Lees AJ, Jenner P, Marsden CD (1986) Basal lipid peroxidation as cause of nigral cell death in Parkinson's disease. Lancet, 2639-40

Dexter D, Carter C, Javoy-Agid F, Lees AJ, Jenner P, Marsden CD (1989a) Basal lipid peroxidation in substantia is nigra increased in Parkinson's disease. J Neurochem 52:53692-697

Dexter DT, Wells FR, Lees AJ, Agid F, Agid Y, Jenner P, Marsden CD (1989b) Increased nigral iron content and alterations in other metal irons occurring in brain in Parkinson's disease. J Neurochem 52:1830-1936

Dexter DT, Carayon A, Vidailhet M, Ruberg M, Agid F, Agid Y, Lees AJ, Wells FR, Jenner P, Marsden CDS (1990) Descreased ferritin levels in brain in Parkinson's disease. J Neurochem 55:16-20

Duvoisin RC (1986) Etiology of Parkinson's disease: current concepts. Clin Neuropharmacol 9 (suppl. 1):S3-S11

Earle KM (1968) Studies on Parkinson's disease including X-ray fluorescent spectroscopy of formalin fixed brain tissue. J Neuropathol Exp Neurol 27:1-14

Graham DG (1979) On the origin and significance of neuromelanin. Arch Path Lab Med 103:359-362

Graham DG, Tiffany SM, Bell WR (1978) Auto-oxidation versus covalent binding of quinones as the mechanism of toxicity of dopamine, 6-hydroxydopamine, and related compounds towards C 1300 neuroblastoma cells in vitro. Mol Pharmacol 14:644-653

Graybiel AM, Hirsch EC, Agid Y (1990) The nigrostriatal system in Parkinson's disease. In: (Streifler MB, Korczyn AD, Melamed E, Youdim MB eds.) 9th International Symposium on Parkinson's disease. Jerusalem, June 5-9. Raven Press New York 53:17-20

Hallgren B, Sourander P (1958) The effect of age on non haemin iron in human brain. J Neurochem 3:41-51

Halliwell B, Gutteridge JMC (1986) Iron and free radical reactions: two aspects of antioxidant protection. Trends Biol Sci 11:1372-1375

Hentze MW, Caugham SW, Roault TA, Barriolanal JG, Dancis A, Harford JB, Klausner RD (1987) Identification of the iron-responsive element for the translational regulation of human ferritin mRNA, Science 238:1570-1573

Hirsch EC, Graybiel AM, Agid Y (1988) Melanized dopaminergic neurons are differentially affected in Parkinson's disease. Nature 334:345-348

Hirsch EC, Graybiel AM, Agid Y (1989) Selective vulnerability of pigmented dopaminergic neurons in Parkinson's disease. Acta Neurol Scand 126:19-22

Hirsch EC, Mouatt A, Graybiel AM, Javoy-Agid F, Agid Y (1991a) The loss of catecholaminergic neurons in Parkinson's disease is heterogeneous in the midbrain. 10th International Symposium on Parkinson's disease, Tokyo
Hirsch EC, Brandel JP, Galleg P, Javoy-Agid F, Agid Y (1991b) Iron and aluminium increase in the substantia nigra of patients with Parkinson's disease: an X-ray microanalysis. J Neurochem 56:446-451
Javoy-Agid F, Hirsch E, Dumas S, Duyckaerts C, Mallet J, Agid Y (1990) Decreased tyrosine hydroxylase messenger mRNA in the surviving dopamine neurons in Parkinson's disease: an in situ hybridization study. Neuroscience 38:245-253
Johnson WG, Hodge SE, Duvoisin R (1990) Twin studies and the genetic of Parkinson's desease – a reappraisal. Mov Dis 5:187-194
Kish SJ, El-Awar M, Schut L, Leach L, Oscar-Berman M, Freedman M (1988) Cognitive dificits in olivopontocerebellar atrophy: Implications for the cholinergic hypothesis of Alzheimer's dementia. Ann Neurol 24:200-206
Koller W, Vetere-Overfield B, Gray RN, Alexander BS, Chin T, Dolezal J, Hassanein R, Tanner C (1990) Environmental risk factors in Parkinson's disease. Neurology 40:1218-1221
Langston JW, Ballard P, Tetrud JW, Irwin I (1983) Chronic parkinsonism in humans due to a product of mepedrine analogue synthesis. Science 219:979-980
Marsden CD (1983) Neuromelanin and Parkinson's disease., J Neural Transm 19(suppl):121-141
Marttila RJ, Lorentz H, Rinne UK (1988) Oxygen toxicity protecting enzymes Parkinson's disease. J Neuro Sci 86:321-331
McGeer PL, Itagaki q, Akiyama H, McGeer EG (1988) Rate of cell death in parkinsonism indicates active neuropathological process. Ann Neurol 24:564-576
Nicklas WJ, Vyas I, Heikkila RE (1985) Inhibition of NADH-linked oxidation in brain mitochondria by MPP^+, a metabolite of the neurotoxin MPTP. Life Science 36:2503-2508
Pall HS, William AC, Blake DR, Luneec J, Gutteridge JM, Hall H, Taylor A (1987) Raised cerebrospinal-fluid copper concentration in Parkinson's disease. Lancet i:238-241
Parker WD, Boyson SJ, Parks JK (1989) Abnormalities of the electron transport chain in idiopathic Parkinson's disease. Ann Neurol 26:719-723
Perry TL, Yong WW (1986) Idiopathic Parkinson's disease, progressive supranuclear palsy and glutathione metabolism in the substantia nigra of patients. Neurosci Lett 67:269-274
Richter C, Park JW, Ames BN (1988) Normal oxidative damage to mitochondrial and nuclear DNA is extensive. Proc Natl Acad Sci USA 85:6465-6467
Riederer P, Rausch WD, Schmidt B, Kruzik P, Konradi C, Sofia E, Danielczyk W, Fisher M, Ogris E (1988) Biochemical fundamentals of Parkinson's disease. Mt Sinai J Med 55:21-28
Saggu H, Cooksey J, Dexter D, Wells FR, Lees A, Jenner P, Marsden CD (1989) A selective increase in particulate superoxide dismutase activity in parkinsonian substantia nigra. J Neurochem 53:692-697
Schapira AHV, Holt IJ, Sweeney M, Harding AE, Jenner P, Marsden CD (1990a) Mitochondrial DNA analysis in Parkinson's disease. Mov Dis 5,4:294-297
Schapira AHV, Cooper JM, Dexter D, Clark JB, Jenner P, Marsden CDS (1990b) Mitochondrial complex I deficiency in Parkinson's disease. J Neurochem 54:823-827
Scheinber IH, Strenlieb I (1984) Wilson's disease. Saunders, Philadelphia
Schermann D, Desnos C, Darchen F, Pollak P, Javoy-Agid F, Agid Y (1989) Striatal dopamine deficiency in Parkinson's disease: role of aging. Ann Neurol 26:551-557
Youdim MBH, Ben-Shachar D (1989) Iron-melanin interaction in substantia nigra as the neurotoxic component of Parkinson's disease. Abstracts from the II international conference on basic and therapeutic strategies for Alzheimer's and Parkinson's disease, Kyoto, Japan, 53

Posttraumatic Epilepsy, Free Radicals and Antioxidant Therapy

A. Mori, M. Hiramatsu, and *I. Yokoi*

Summary

Each year, head injuries from car accidents and other causes result in a large population of patients with epileptic seizure disorders. The development of such posttraumatic epilepsy may be related to the breakdown of red blood cells and hemoglobin within the central nervous system. Subpial injection of hemoglobin or iron salts into rat cortex is known to induce a chronic epileptic focus. We observed the formation of superoxide anion and hydroxyl radical (•OH) after ferric chloride injection into rat cerebral cortex, and suggested that these radicals, especially •OH, may be responsible for the initiation of lipid peroxidation in neuronal membranes and for the accelerated production of endogenous convulsants, i. e., guanidino compounds, in the brain. These disorders may in turn lead to epileptogenicity. We found that treatment with epigallocatechin or a phosphate diester of vitamins E and C, which are potent •OH scavengers, significantly inhibited the formation of thiobarbituric acid-reactive substances and epileptic discharges in the iron-induced epileptic focus.

Introduction

Posttraumatic epilepsy is characterized by epileptic seizures due to brain damage secondary to head injury. It has been recognized since the days of Hippocrates, but the pathogenesis of posttraumatic epilepsy is still unclear. From the standpoint of the first occurrence, neurosurgeons classify posttraumatic epilepsies into two types, i. e., early epilepsy (or early seizures) and late epilepsy. Early epilepsy is defined as a convulsion occurring within 1 week of head trauma. Seizures occur more frequently during the first week after injury than during the next 7 weeks, and patients with early epilepsy have a high risk of late epilepsy. During the 1980s in the USA, motor vehicle accidents caused approximately 500000 head injuries annually (Kalsbeck et al. 1980), and the number of these injuries is likely to increase. The occurrence of posttraumatic epilepsy has been reported to be 15% in patients with head injury (Guidice and Berchou 1987) and was reported to be up to 45% in cases of penetrating head injuries (Ruggles et al. 1988). In Japan, the occurrence of posttraumatic epilepsy is thought to be approximately 150000 cases annually (Japanese population equal about 120 million), which is equal to about 10% of all patients hospitalized with head injuries (Manaka 1980). Offering a possible mechanism for the development of posttraumatic epilepsy, computerized tomography (CT) scanning studies have demonstrated that the main factor in the development of early and late epilepsy is focal hemorrhagic brain damage (d'Alessandro et al. 1982).

Using iron-induced epileptic rats (Willmore et al. 1978a, b, c), we have been working on the pathogenesis of posttraumatic epilepsy, especially on the relationship between the generation of free radicals and the development of posttraumatic epilepsy and also on the rational treatments for it, based on these experimental results. Here, we introduce some topics arising mainly from investigations in our laboratory.

First we discuss focal epileptic discharges induced by injection of iron into rat cerebral cortex, which is an experimental model of posttraumatic epilepsy.

Iron-Induced Epileptic Discharges in Rat: An Experimental Model of Posttraumatic Epilepsy

Willmore et al. (1978a, b, c), first reported that a single injection of 5 or 10μl of ferrous or ferric chloride into rat or cat sensorimotor cortex resulted in chronic recurrent focal paroxysmal electroencephalographic discharges as well as behavioral convulsions.

The involvement of iron-mediated oxygen free radical species and neuronal lipid peroxidation in the iron-induced seizure mechanism has been suggested from the first of these studies. Khochi (1983) observed that epileptic discharges were induced 15 min after 0. 1 *M* $FeCl_3$ was injected into the rat sensorimotor cortex, and continued for more than 5 months after the injection. Figure 1 shows electrocorticography tracings from a rat injected with 5μl of 100 m*M* of $FeCl_3$ into sensorimotor cortex. Spike discharges were observed 15 min after the injection, and developed into epileptiform discharges thereafter.

Rosen and Frumin (1979) reported that hemoglobin could induce epileptic discharges just as the iron ion does. This evidence suggests that the epileptogenic effect of hemoglobin may depend on ionic iron released from hemoglobin. Otherwise, hemoglobin itself may also be epileptogenic, because free hemoglobin promotes hydroxyl radical (•OH) and thiobarbituric acid-reactive substances (TBARS) formation, an index of lipid peroxidation, as a biological Fenton reagent (Sadrzadeh et al. 1984).

Generation of Oxygen Free Radical Species by Injection with Ferric or Ferrous Ion into Rat Cerebral Cortex

The addition of iron salts or heme compounds to tissue suspensions or to solutions of polyunsaturated fatty acids results in the formation of superoxide anion (O_2^-), •OH and peroxide radicals (O'Brien 1969; Wever et al. 1973; Wills 1965; Aisen 1977). Therefore, it has been proposed that iron-induced formation

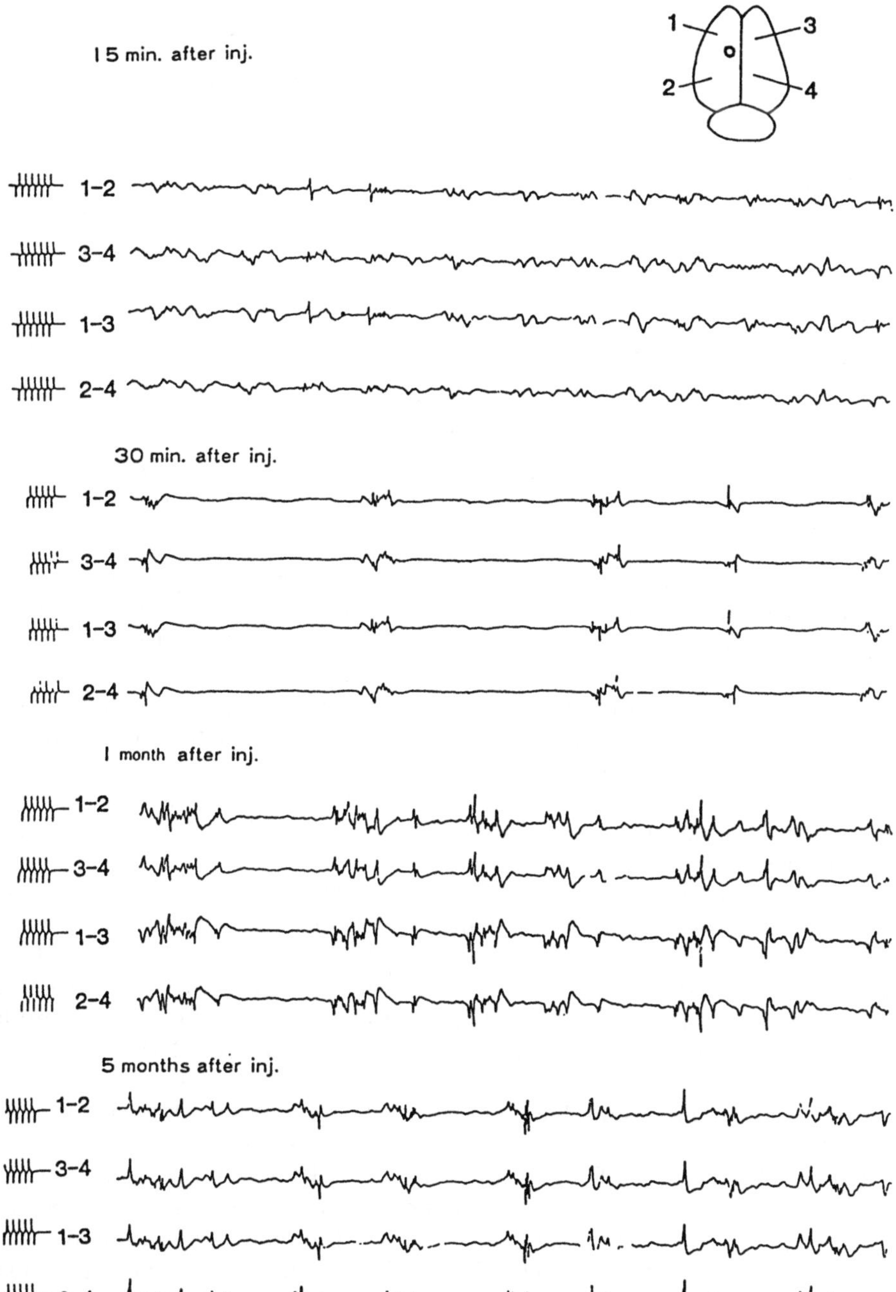

Fig. 1. Effect of Fe^{3+} on rat electrocorticogram. Sporadic spikes were induced in the Fe^{3+} injection site 15 min after the injection and developed into diffuse multiple spike discharges until 30 min after. Then recurrent epileptic discharges were recorded even at 5 months after the injection. ○, Fe^{3+} injected position, *1, 2, 3* and *4*, positions of electrodes

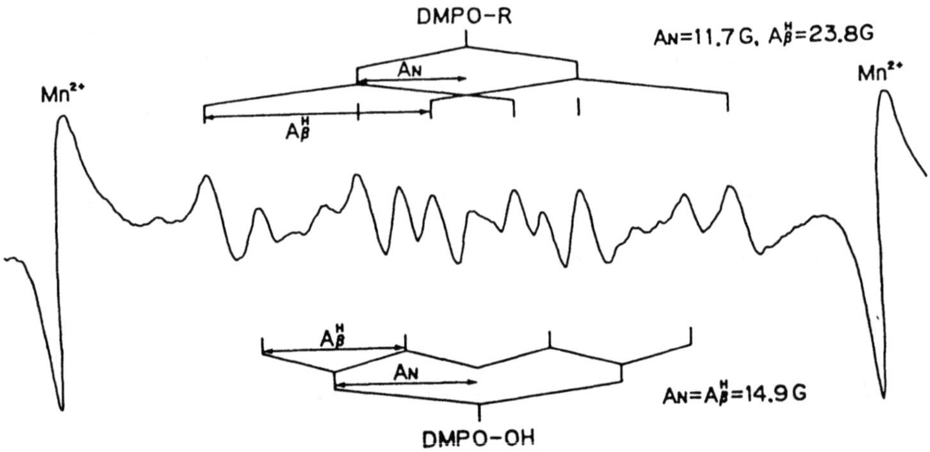

Fig. 2. Generation of hydroxyl radical (•OH) and C-centered radical (•R) 30 min after injection with iron ion into rat brain, determined as DMPO spin adduct by an ESR spectrometer

of free radicals and powerful oxidants causes neural lipid peroxidation with development of an epileptic focus. We first demonstrated transient formation of oxygen free radicals after the injection of iron salt into the rat cerebral cortex (Willmore et al. 1983; Hiramatsu et al. 1984).

Figure 2 shows the formation of •OH, hydrogen (•H) and C-centered (•R) radicals 30 min after the intracortical injection of 5μl of 100 m*M* ferric citrate, as demonstrated by electron spin resonance (ESR) spectrometry and spin-trapping methods using 5, 5-dimethyl-1-pyrroline-1-oxide (DMPO; Ueno et al. 1984). Active generation of free radicals, especially •OH, was observed in the iron-injected side of rat cerebral cortex.

On the other hand, we observed experimentally that ferric as well as ferrous salts could produce free radicals, mainly •OH, in single water solution without hydrogen peroxide (Mori et al. 1989b). This reaction is thought to be a different reaction from a Fenton-type reaction, in which hydrogen peroxide is essential. This evidence suggests that not only ferric ion but also ferrous ion can produce oxygen free radicals even in brain tissue, independently of the existence of hydrogen peroxide.

Concerning hemoglobin, Sadrzadeh et al. (1984) reported that free hemoglobin catalyzes the formation of •OH. The mechanism is very likely a Fenton-type reaction, wherein heme ferric ion decomposes H_2O_2 to form methemoglobin, •OH and OH^-. We also observed that hemoglobin markedly generated •OH on addition of H_2O_2, both at pH 4.0 and pH 7.4 (Mori et al. 1989b). Figure 3 shows the ESR spectra of hydroxyl (DMPO-OH) and C-centered spin adducts (DMPO-R) detected in hemoglobin solution. Free hemoglobin is thought to be potentially hazardous, i. e., hemoglobin promotes the oxidation and destruction of a variety of biomolecules (Sadrzadeh et al. 1984; Gutteridge 1986) and may result in cellular injury. Therefore, not only free iron released from hemoglobin but also hemoglobin itself could be responsible for the pathogenesis of posttraumatic epilepsy, as mentioned above.

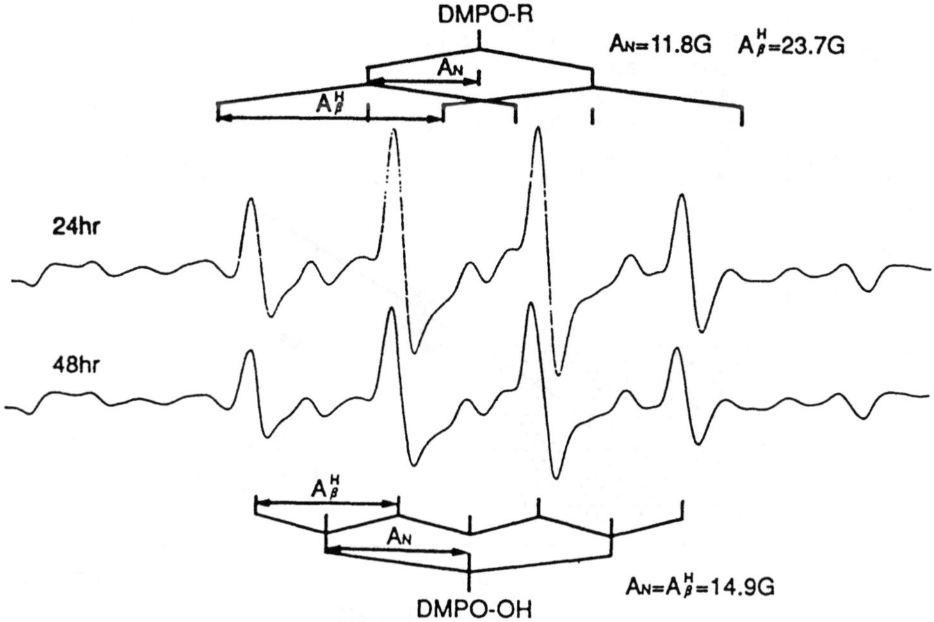

Fig. 3. Generation of hydroxyl radical (•OH) and C-centered radical (•R) in hemoglobin-H_2O_2 solution. Hemoglobin (200 μg) was dissolved in 20 ml of 100 m*M* Pipes buffer (pH7.4) and 20 μl H_2O_2 (1.37 *M*) and 20 μl of DMPO added to 180 μl of this solution. Two min after stirring the free radicals were analyzed as DMPO adducts by an ESR spectrometer

Peroxidation of Membrane Lipid by Oxygen Free Radicals

Neural cell membranes contain rich unsaturated fatty acids in their structure and are peroxidized easily by oxygen free radicals. This causes damage in the neural cell membrane and thereby produces loss of fluidity and breakdown of the membrane functions and transmembrane ionic gradients. Willmore and Rubin (1982) demonstrated significant formation of TBARS, an index of lipid peroxidation, after injection of $FeCl_2$ into rat isocortex. We also observed increased levels of TBARS in the brain after injection with $FeCl_3$, and found that there is a significant positive relationship between •C and TBARS level in ferric ion-induced epileptic focus (Mori and Hiramatsu 1984) (Fig. 4). Increased products of lipid peroxidation were also measured by fluorescence in the dissected hippocampus injected with $FeCl_2$ (Willmore et al. 1986).

Neurotransmitter Disorders Induced by Iron Ion

Glutamic acid and aspartic acid are thought to be principal excitatory neurotransmitters and γ-aminobutyric acid (GABA) to be a principal inhibitory neurotransmitter. We analyzed amino acid levels in the brain, following the

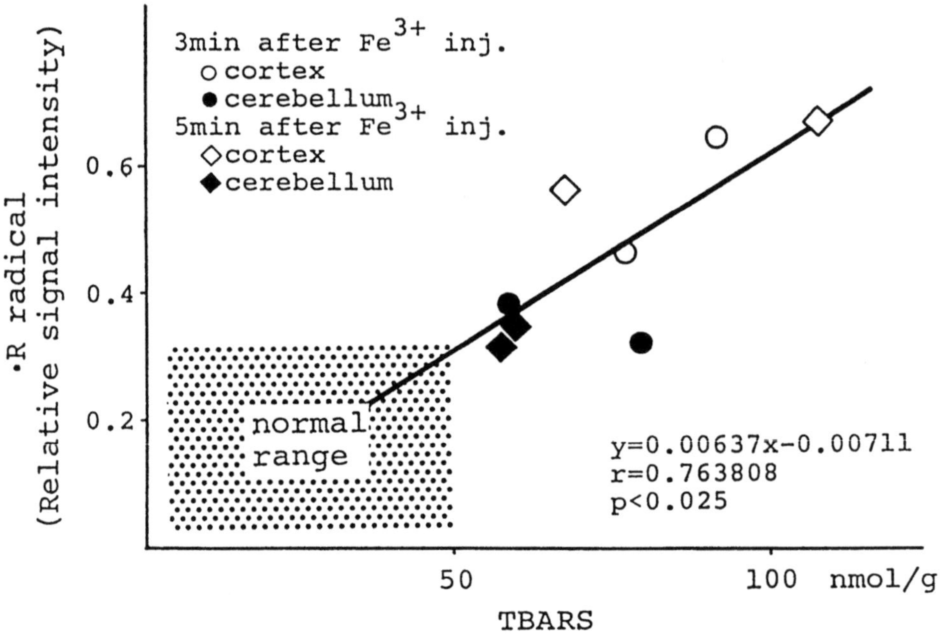

Fig. 4. C-centered radical (•R) and TBARS levels in rat brain after injection with iron ion. ○, cortex and ●, cerebellum 3 min after Fe^{3+} injection; ◇, cortex and ◆ cerebellum 5 min after Fe^{3+} injection

injection of ferric chloride into the cerebral cortex, using high-performance liquid chromatography (HPLC) and observed almost no significant change in amino acid neurotransmitter in the brain 2 months after the injection, although aspartic acid, glutamic acid and also GABA levels decreased temporarily within 48 h after the injection (Hiramatsu et al. 1987; Shiota et al. 1989). Zhang et al. (1987) reported that GABA uptake and release were both significantly decreased in synaptosomes from rats with ferrous-induced seizures and that GABA release was more severely damaged than uptake, suggesting that this imbalance between release and uptake may reduce the GABAergic inhibition, which may contribute to iron-induced seizure mechanism.

On the other hand, we studied release of aspartic acid and glutamic acid from slices of cerebral cortex and found increased release of aspartic acid in the acute stage, i.e., 30 min after the injection of ferric chloride, though no significant difference was observed in the chronic stage, i.e., 3 weeks after the injection (Janjua et al. 1990). These facts suggest that the increased release of excitatory amino acid, i.e., aspartic acid, and the decreased release of inhibitory amino acid, i.e., GABA, may relate to the seizure mechanism of so-called early epilepsy and also may relate to the development of late epilepsy.

Monoamines such as dopamine, noradrenaline and serotonin are thought to be inhibitory neurotransmitters or neuromodulators. We observed decreased levels of dopamine in some parts of the brain, such as the hypothalamus, midbrain, hippocampus and cortex, and also a decreased level of noradrenaline in cerebellum in the acute stage of ferric ion-induced epileptic focus tissue (Hiramatsu et al. 1987). We also demonstrated accelerated release of dopamine from striatum in acute stage after the injection of ferric chloride, utilizing a brain dialysis method (Yokoi et al. 1990). This finding suggests that the formation of an epileptic focus induced by iron ions accompanies the activation of dopaminergic neurons. Monoaminergic neurotransmitters may also play some role in the mechanism of "early epilepsy" and in the procedure of chronic epileptic focus formation.

Accelerated Formation of Guanidino Compounds (Endogenous Convulsants) by Iron Ion

Some guanidino compounds are known to be endogeneous convulsants (Mori 1987). Methylguanidine, a simple guanidino compound and an endogeneous potent convulsant (Matsumoto et al. 1976), is shown to be produced from creatinine by iron-catalized free radical reaction (Nagase et al. 1986). We observed that brain levels of methylguanidine and guanidinoacetic acid fluctuated greatly after the ferric chloride injection and recovered transitorily to normal levels 24-48 h after the injection (Fukushima 1987). These substances finally increased to two or three times normal levels 2 months after the injection, at which time chronic epileptogenic foci had formed (Fig. 5). Such long-lasting increased levels of methylguanidine and guanidinoacetic acid were observed not only in the chronic epileptogenic focus induced by iron injection but also in the amygdala kindled rat brain (Hirayasu et al. 1989). These findings suggest that guanidino compund may be related to excitatory mechanisms of epileptogenesis in "late epilepsy."

On the other hand, we found that some guanidino compounds generate •OH and •C (Mori 1987). For example, α-guanidinoglutaric acid generates •OH and •C in extremely acidic and basic solutions. These facts suggest that a part of the neurotoxicity of guanidino compound may be based on peroxidation of polyunsaturated fatty acids in neuronal membranes by such peroxide radicals generated from guanidino compounds.

Is Oxygen Inhalation an Accelerator of Posttraumatic Epileptic Focus Formation?

Oxygen inhalation is an emergency measure routinely used for patients with head injury, although a high concentration of oxygen is reported to be hazardous in biological systems (Fracica et al. 1991). We studied the effect of 100% oxygen on free radical generation, TBARS production and superoxide dismutase (SOD) activity in the rat brain 1, 2 and 3 weeks after injection with $FeCl_3$. The results are as follows (Fig. 6):

1. Oxygen inhalation accelerated •OH generation insignificantly 1 week after but significantly 3 weeks after the injection.

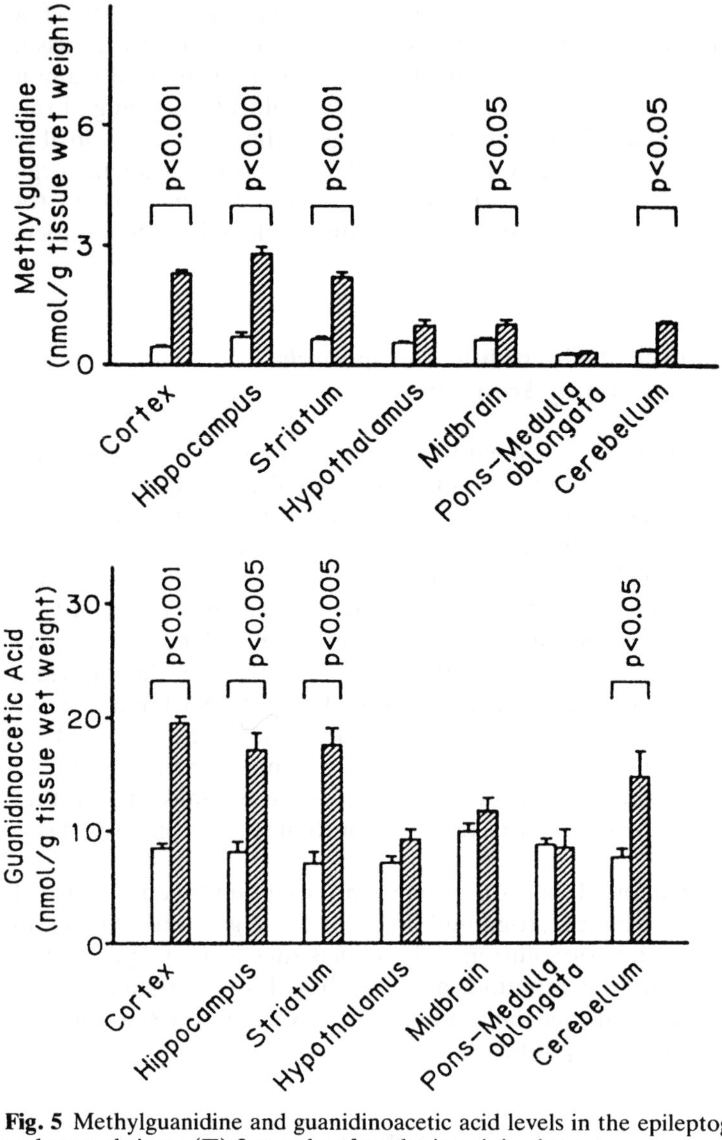

Fig. 5 Methylguanidine and guanidinoacetic acid levels in the epileptogenic focus tissue (▨) and control tissue (□) 2 months after the iron injection

2. Oxygen inhalation accelerated •R generation 1 week after, inhibited it 2 weeks after and re-accelerated it 3 weeks after the injection.
3. Oxygen inhalation accelerated TBARS production 1 week after the injection and inhibited it 3 weeks after the injection.
4. Oxygen inhalation accelerated mitochondrial SOD activity 1 week after and inhibited it 3 weeks after the injection, while inhibiting postmicrosomal, i. e., cytosolic, SOD activity 1 week after and accelerating it 3 weeks after the injection.

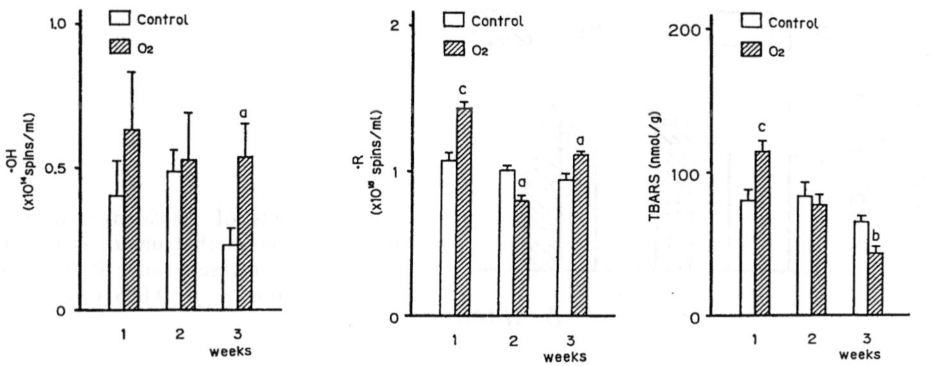

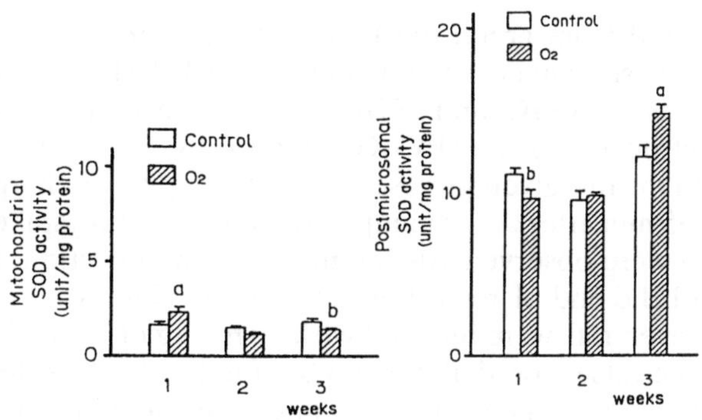

Fig. 6. Effects of 100% oxygen inhalation on hydroxyl radical (•OH) C-centered radicals (•R), thiobarbituric acid-reactive substances (TBARS) levels and superoxide dismutase (SOD) activity in rat brain. The values are the means ± SEM of 5-10 rats. *a*, $p<0.05$ vs control; b, $p<0.01$ vs control; *c*, $p<0.001$ vs control

These findings suggest that prolonged treatment with high concentrations of oxygen accelerated free radical generation by iron injection and increased lipid peroxidation in the brain. SOD activities change in a biphasic manner during oxygen inhalation. Increased SOD activities, i. e., mitochondrial 1 week after the injection and postmicrosomal 3 weeks after, may relate to a compensatory mechanism against the increased oxygen free radicals. On the other hand, we observed significantly increased levels of methylguanidine in the brain 3 weeks after injection with iron ion (Fig. 7). Oxygen inhalation is a clinical first aid for patients with severe head injury, but the clinically beneficial concentration and period for treatment should be reexamined carefully from the viewpoint of oxygen intoxication.

Antioxidant Therapy for Posttraumatic Epilepsy: Experimental Studies

Willmore and Rubin (1981, 1984) reported that pretreatment with α-tocopherol prevented the development of iron-induced epileptiform activity and accelerated the resolution of TBARS formation within the iron-injected rat cortex.

(-)-Epigallocatechin (EGC) and (-)-epigallocatechin-3-0-gallate (EGCG)

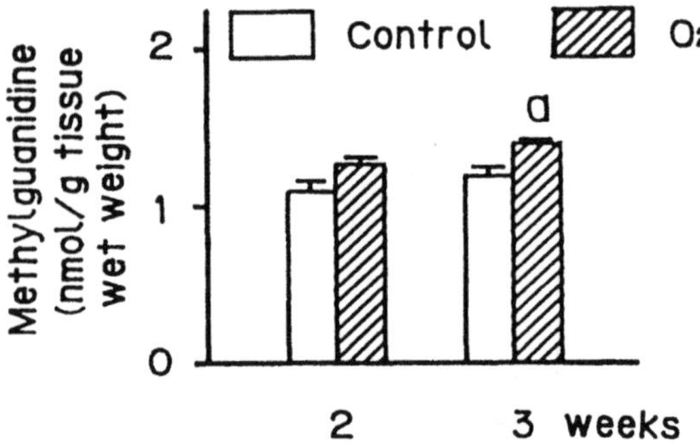

Fig. 7. Effects of 100% oxygen (▩) inhalation on methylguanidine in rat brain. The values are mean ± SEM of 5-6 determinations. *a*, $p<0.005$ vs control (□)

isolated from green tea leaves, *Rhei rhizoma*, are known to be oxygen free radical scavengers (Nonaka et al. 1983; Uchida et al. 1987). We examined the effects of EGC and EGCG on the occurrence of epileptic discharges in electrocorticogram (ECoG) induced by iron injection into the rat brain.

In the rats given sucrose as a control, spike discharges (SD) appeared in ECoG 15-55 min after the iron injection, and then polyspikes (PS) and/or ictal patterns (IP) were observed 70-100 min after the injection. Epileptic discharges lasted until the end of recordings at 210 min (Fig. 8).

Seven rats were injected with 50 mg/kg of EGC (i. v.) after the iron injection. Three rats showed almost no SD for 4 h. In the other three rats, IP started 120 min or 150 min after the iron injection. No inhibitory effect was observed in the last rat (Fig. 8).

Four rats were injected with EGCG i. p. prior to the iron injection. No SD were induced in two rats. In the other two rats, SD and PS started 55 min and 150 min after the iron injection, but no rat showed IP (Yokoi et al. 1989).

L-Ascorbic acid 2-[3, 4-dihydro-2, 5, 7, 8-tetramethyl-2-(4, 8, 12-trimethyl-tridecyl)-2H-1-benzopyran-6-yl-hydrogen phosphate] (EPC) is a phosphate diester of vitamin E and vitamin C (Ogata 1986). We found EPC to be a potent scavenger of •OH generated by the iron-catalyzed reaction, and that EPC (60 mg/kg, i. p., 30 min before iron injection) significantly inhibited the TBARS formation in the iron-induced epileptic focus (Mori et al. 1989a). Next we examined the effect of EPC on the induction of seizure discharges in the rat ECoG induced by iron and observed that, when 500 nmol of ferric chloride together with 500 pmol of EPC (in 5μl) was injected into the rat motor cortex, epileptic discharges were not induced (Fig. 9). ED_{50} was around 250 pmol. On the other hand, when 70 nmol of EPC was injected intracerebroventricularly 60 min prior to the iron injection into cortex, ictal seizures were not induced. When 14 nmol of EPC was injected beforehand, the occurrence of seizure discharges was suppressed in 50% of rats.

Up to the present, most clinical antiepileptics known to inhibit the propagation of epileptic discharges in the brain have not been very rational treatments for epileptogenecity. Along with the elucidation of a seizure mechanism, the rational treatment of epilepsy has become a big problem awaiting a solution, such as therapy for epilepsy by excitatory amino acid receptor antagonists and

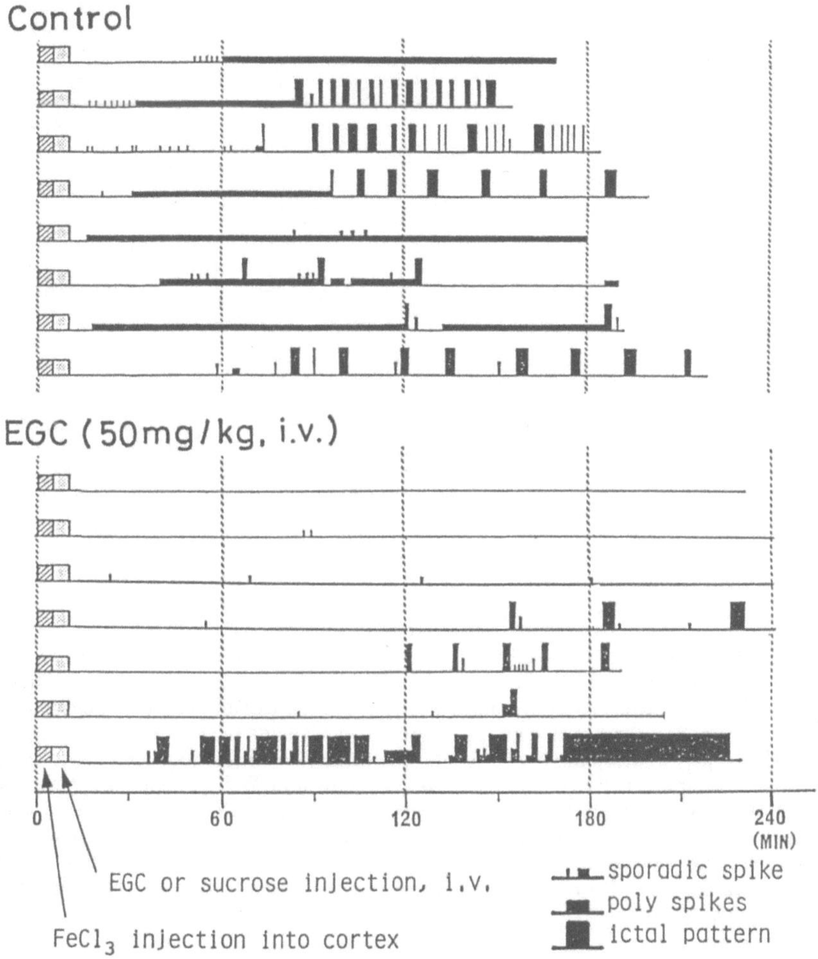

Fig. 8. Effect of (-)-epigallocatechin (EGC) on the development of seizure activity induced by $FeCl_3$. $FeCl_3$ saline solution (100 mM, 5 μl) was injected into the sensorimotor cortex of rats. Then either EGC (50 mg/kg) or sucrose (50 mg/kg) was injected i. v. to study the effects of EGC on the occurrence of epileptic discharges in electrocorticograms induced by $FeCl_3$

inhibitory amino acid receptor agonists, or receptor modulators. Prevention and/or treatment of posttraumatic epilepsy by anti-oxidants, e. g., α-tocopherol, EGC and EPC, could be a possible rational treatment, especially as prophylactics for "early epilepsy."

Conclusion

Head injury or hemorrhagic cortical infarction results in extravasation of blood and the breakdown of red blood cells and hemoglobin. Biological iron is

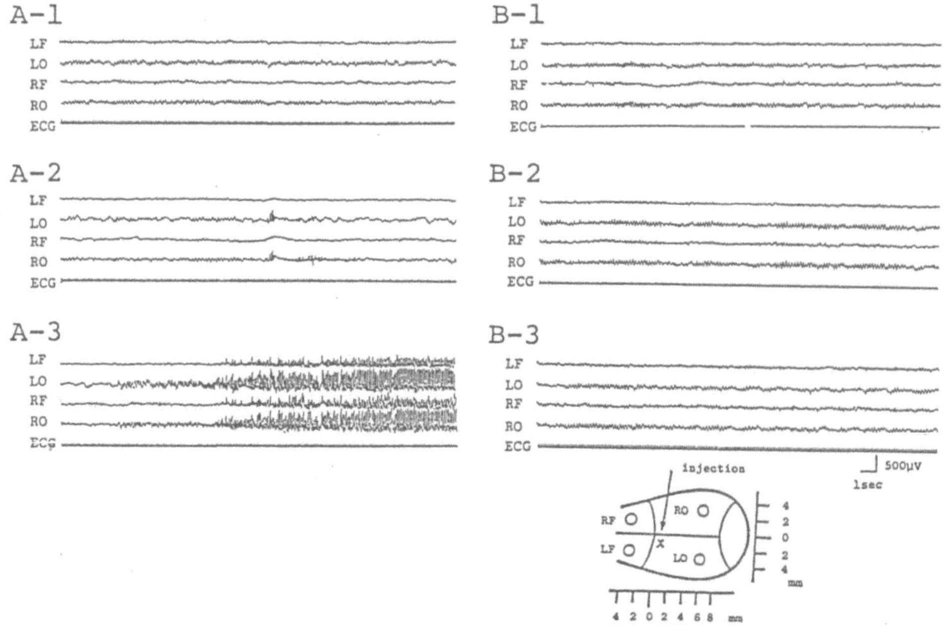

Fig. 9 A, B Effect of a phosphate diester of vitamins E and C (EPC) on the development of seizure activity induced by $FeCl_3$. **A-1** 5 μl of $FeCl_3$ solution (100 m*M*) was injected into the sensorimotor cortex; **A-2** polyspikes were seen in the LO and RO (see below) 62 min after the injection; **A-3** ictal pattern began to be seen from all leads 175 min after the injection. **B-1** 5 μl of EPC (100 μM) solution dissolved by $FeCl_3$ solution (100 m*M*) was injected into the sensorimotor cortex, no spike activity was seen; **B-2** 60 min and **B-3** 180 min after the injection. The recordings were from four epidural electrodes: *LF,* unipolar recording from left frontal electrode; *LO,* unipolar recording from left occipital electrode; *RF,* unipolar recording from right frontal electrode; and *RO,* unipolar recording from right occipital electrode. *ECG,* electrocardiogram. Positions of each electrode and the injection site of drugs are shown in the *lower right corner.*

normally protein-bound in hemoglobin and transferrin. Iron liberated from hemoglobin is thought to be associated with the generation of active oxygen species. Moreover, hemoglobin itself may promote oxygen free radical generation. Oxygen free radicals, especially •OH, are responsible for the induction of peroxidation of neuronal lipids, that is, an injury of neuronal membranes. On the other hand, •OH accelerates production of guanidino compounds in the brain, such as methylguanidine and guanidinoacetic acid. They are endogenous convulsants in the brain. These reactions may be followed by excitatory and inhibitory neurotransmitter disorders, which may lead to the development of epileptic discharges in the epileptogenic focus. Prevention and/or treatment of posttraumatic epilepsy by antioxidants, e. g., α-tocopherol, EGC, and EPC, could become a possible rational treatment, especially as prophylactics for early epilepsy.

Acknowledgments. This study was supported by Grant in Aid from the Ministry of Education, Science and Culture of the Japan Government. We gratefully thanks Mr. H. Kabuto and Miss. R. Edamatsu for their technical assistance.

References

Aisen P (1977) Some physiochemical aspects of iron metabolism. Ciba Found Symp 51:1-14

d'Alessandro R, Tinuper P, Ferrara R, Cortelli P, Pazzaglia P, Sabattini L, Frank G, Lugaresi E (1982) CTScan prediction of late posttraumatic epilepsy. J Neurol Neurosurg Psychiatry 45:1153-1155

Fracica PJ, Piantadosi CA, Crapo JD (1991) Oxygen toxicity. Lung: Scientific Foundations Vol. 2, pp 2155-2161

Fukushima M (1987) Guanidino copounds in iron-induced epileptogenic foci of rats. Okayama Igakkai Zasshi 99:1-20 (in Japanese with English abstract)

Guidice MA, Berchou RC (1987) Post-traumatic epilepsy following head injury. Brain Inj 1:61-64

Gutteridge JMC (1986) Iron promoters of the Fenton reaction and lipid peroxidation can be released from hemoglobin by peroxides. FEBS Lett 201:291-295

Hiramatsu M, Mori A. Kohno M (1984) Formation of peroxyl radical after $FeCl_3$ injection into rat isocortex. Neurosciences 10:281-284

Hiramatsu M, Fukushima M, Kabuto H, Edamatsu R, Mori A (1987) Amino acids, 5-hydroxytryptamine, and catecholamines in iron-induced epileptogenic foci of the rat. Adv Epileptol 16:151-154

Hirayasu Y, Morimoto K, Otsuki S. Mori A (1989) Regional changes of guanidino compounds in the amygdala kindled rat brain. American Epilepsy Society Abstract

Janjua NA, Mori A, Hiramatsu M (1990) Increased aspartic acid release from the iron-induced epileptogenic focus. Epilepsy Res 6:215-220

Kalsbeck WD, McLaurin RL, Harris BSH, Miller JD (1980) Report on the national head and spinal cord survey. J Neurosurg 53:519-531

Khochi H (1983) Brain active oxygen, free radicals, lipid peroxide and the redox state of glutathione in the Fe^{3+} induced epileptic focus of rat. Okayama Igakkai Zasshi 95:271-282 (in Japanese with English abstract)

Manaka S (1980) Post-traumatic epilepsy. Geka Mook 11:195-206 (in Japanese)

Matsumoto M, Kobayashi K, Kishikawa H, Mori A (1976) Convulsive activity of methylguanidine in cats and rabbits. IRCS Med Sci 4:65

Mori A (1987) Biochemistry and neurotoxicology of guanidino compounds. History and recent advances. Pavlovian J Biol Sci 22:85-94

Mori A, Hiramatsu M (1984) A role of radical reaction on the neuronal membrane in the seizure mechanism. In: Annual report of research. Cause and treatment of refractory epilepsies. pp 175-179 (in Japanese with English abstract)

Mori A, Edamatsu R, Kohno M, Ohmori S (1989a) A new hydroxyl radical scavenger: EPC-K_1. Neurosciences 15:371-376

Mori A, Hiramatsu M, Edamatsu R, Kohno M (1989b) Possible involvement of oxygen free radicals in the pathogenesis of post-traumatic epilepsy. In: Hayashi O et al. (eds) Medical, biochemical and chemical aspects of free radicals. Elsevier, Amsterdam, pp 1249-1252

Nagase S, Aoyagi K, Narita M, Tojo S (1986) Active oxygen in methylguanidine synthesis. Nephron 44:299-306

Nonaka G, Kashiwahara O, Nishioka I (1983) Tannins and related compounds XV. Chem Pharm Bull (Tokyo) 31:3906-3914

O'Brien PJ (1969) Intracellular mechanisms for the decomposition of a lipid peroxide by metal ions, heme compounds and nucleophiles. Can J Biochem 47:485-492

Ogata K (1986) Phosphoric acid diesters or their salts and process for producing the same. US Patent No 4564, 686

Rosen AD, Frumin NV (1979) Focal epileptogenesis after intracortical hemoglobin injection. Exp Neurol 66:277-284

Ruggles KH, Hansotia PL, Ahmann PA (1988) Head trauma and epilepsy. Wis Med J 87:16-18

Sadrzadeh SMH, Panter SS, Hallaway PE, Eaton JW (1984) Hemoglobin – A biological Fenton reagent. J Biol Chem 259:14354-14356

Shiota A, Hiramatsu M, Mori A (1989) Amino acid neurotransmitters in iron induced epileptic foci of rats. Res Commun Chem Pathol Pharmacol 66:123-133

Uchida S, Edamatsu R, Hiramatsu M, Mori A, Nonaka G, Niwa M, Ozaki M (1987) Condensed tannins scavenge active oxygen free radicals. Med Sci Res 15:831-832

Ueno I, Kohno M, Yasuhisa K, Hirano I (1984) Quantitative determination of the superoxide radicals in the xanthine oxidase reaction by measurement of the electron spin resonance signal of the superoxide radical spin adduct of 5, 5-dimethylpyrroline-1-oxide. J Pharmaco Dynam 7:563-569

Wever R, Qudega B, Van Gelder BF (1973) Generation of superoxide radicals during the autoxidation of mammalian oxyhemoglobin. Biochim Biophys Acta 302:475-478

Willmore LJ, Rubin JJ (1981) Antiperoxidant pretreatment and iron-induced epileptiform discharges in the rat: EEG and histopathologic studies. Neurology 31:63-69

Willmore LJ, Rubin JJ (1982) Formation of malonaldehyde and focal brain edema induced by subpial injection of $FeCl_2$ into rat isocortex. Brain Res 246:113-119

Willmore LJ, Rubin JJ (1984) Effects of antiperoxidants on $FeCl_2$ induced lipid peroxidation and focal edema in rat brain. Exp Neurol 82:62-70

Willmore LJ, Sypert GW, Munson JB, Hurd RW (1978a) Chronic focal epileptiform discharges induced by injection of iron into rat and cat cortex. Science 200:150-153

Willmore LJ, Hura RW, Sypert GW (1978b) Epileptiform activity initiated by pial iontophoresis of ferrous and ferric chloride on rat cerebral cortex. Brain Res 152:406-410

Willmore LJ, Sypert GW, Munson JB (1978c) Recurrent seizures induced by cortical iron injection: a model of post-traumatic epilepsy. Ann Neurol 4:329-336

Willmore LJ, Hiramatsu M, Khochi H, Mori A (1983) Formation of superoxide radicals after $FeCl_3$ injection into rat isocortex. Brain Res 277:293-296

Willmore LJ, Triggs WJ, Gray JD (1986) The role of iron-induced hippocampal peroxidation in acute epileptogenesis. Brain Res 382:422-426

Wills ED (1965) Mechanism of lipid peroxide formation in tissues. Role of metals and haematin proteins in the catalysis of the oxidation of unsaturated fatty acid. Biochim Biophys Acta 98:238-251

Yokoi I, Kabuto H, Akiyama K, Mori A, Ozaki M (1989) Tannins inhibit the occurrence of epileptic focus induced by $FeCl_3$ injection in rat. Jpn J Psychiatry Neurol 43:552-553

Yokoi I, Kabuto H, Yamamoto M, Mori A (1990) Effect of ferric ion and active oxygen scavenger on monoamine release in the striatum of the rat. Jpn J Psychiatry Neurol 44:440-441

Zhang ZH, Zuo CH, Wu XR, Hua Y, Wang H (1987) The effect of lipid peroxidation on GABA uptake and release in iron-induced seizures. Chin Med J 102:24-27

Lipid Peroxidation in the Pathology of the Retina

M. Doly and *M.-T. Droy-Lefaix*

Summary

The major origin of lipid peroxidation in vivo is the production of oxygen free radicals by inflammatory cells. These reactive species attack phospholipids, disorganize cell membranes and thus cause tissue injury. Among the neurosensory tissues, the retina is particularly rich in unsaturated fatty acids and is thus highly sensitive to oxygen free radicals. Lipid peroxidation in this tissue is extremely deleterious, because the electroretinogram (ERG) is produced by a series of differences in membrane ionic permeabilities. Retinal pathologies associated with overproduction of free radicals can be classed into two main groups.

The first group includes pathologies directly resulting from free radical damage, such as:

1. retinal photodegeneration caused by intense visible light or UV illumination and associated with the photolysis of water or other essential molecules;
2. retinopathy of prematurity due to free radical damage to Müller cells implicated in the ERG genesis; and
3. ischemia-reperfusion, which induces an excessive production of free radicals in the reperfusion phase and constitutes one of the major ocular diseases.

The second group of pathologies corresponds to inflammatory conditions in which free radicals are indirectly implicated, including:

(a) auto-immune uveitis, characterized by infiltration of inflammatory cells generating free radicals;
(b) diabetic retinopathy, characterized by a thickening of the basal membrane, a rupture of the blood-retinal barrier and an abnormally high level of membrane lipid oxidation; and
(c) xenobiotic-induced retinopathies, such as that induced by chloroquine, in which free radical production is associated with the presence of platelet-activating factor (PAF), a potent activator of leukocytes.

In conclusion, there is considerable evidence from experimental studies that oxygen free radicals are implicated in the pathology of diverse retinal diseases. The therapeutic potential of free radical scavengers is currently being evaluated. Encouraging preliminary results have been obtained in studies where these agents have been used preventively.

Introduction

The retina is the neuro-sensorial tissue of the eye in which visual information is generated. Retinal tissue is well differentiated and encloses a large variety of nervous cells. The retinal function schematically consists in the transformation of an optical image into a nervous signal. Such a process needs the intervention of a transduction mechanism. In retina, this system is relatively complex, with precise sequences of physico-chemical events.

The organization of functional retina is classically described according to the choroido-vitreous axis. Three types of cells are present:

1. photoreceptors known to generate visual information;
2. bipolar cells, a variety of pure intraretinal neurons which corresponds to the inner nuclear layer; and
3. ganglion cells, a third variety of neurons with their axons patterned to constitute the optic nerve.

Two internervous areas, called plexiform layers, parallel this functional organization: the outer plexiform layer, a synaptic area between photoreceptor and bipolar cell, and the inner plexiform layer, a synaptic area between bipolar and ganglion cell. Along this choroido-vitreous axis, three types of neurons with lateral disposition have been shown to modulate transport of the visual information: horizontal cells, at the level of the outer plexiform layer; amacrine cells, at the level of inner plexiform layer; and interplexiform cells, at the level of both plexiform layers. To carry visual information, an initial sensorial signal takes place exclusively in photoreceptor cells, where light photons can be detected. On the other hand, other associated neurons participate in the elaboration and modulation of the signal received.

The different steps of phototransduction, localized in the rod, are now well elucidated. Firsty at the level of the saccular membrane of the photoreceptor outer segment, light photons are absorbed by a photopigment, rhodopsin. This absorption is detected by linkage with a G-protein, called transducin. This rhodopsin photoisomerization induces transducin activation. This reaction needs an intracellular effector, the phosphodiesterase (PDE), which is known to hydrolyze cyclic-GMP (c-GMP) into 5′-GMP. Rhodopsin isomerization provokes, into the rod outer segment, a decrease of cGMP concentration. It is now well established that Na^+ permeability of the outer segment membrane is cGMP-dependent. In these conditions, rhodopsin isomerization, through variation of cGMP concentration, provokes the closing of the rod outer segment membrane Na^+ channels and, consequently, hyperpolarization. Such a variation of membrane potential initiates a specific electrophysiological signal. The phototransduction mechanism involves two membranes: the saccular membrane and the rod plasma membrane. This initial signal, generated by the photoreceptor, is transmitted to various excitable retina cells which also generate a specific electrophysiological response. The summation of these individual responses determines the receptor potential of the retina (the electroretinogram ERG). As the retinal membranes are rich in polyunsaturated lipids, they are particularly sensitive to oxygenated free radicals known to induce deleterious effects.

Oxygenated Free Radicals

Physiologically, and in some pathological circumstances such as inflammation, the retina can produce oxygenated free radicals (Mittag 1984; Freeman and Crapo 1983). On their outermost orbital, these free radicals have an impaired electron and are therefore very reactive. Primary oxidants are superoxide anion ($O_2^{\bullet}$), hydrogen peroxide (H_2O_2), and hydroxyl radical ($OH^{\bullet}$). Secondary oxidants are the lipoperoxides (Fridovich 1976). $O_2^{\bullet -}$, a weak oxidant in aqueous solution, plays a major role as an $OH^{\bullet}$ precursor (Fridovich 1983). However, H_2O_2 without an impaired electron, is a strong oxidant. In the presence of ferrous ions (Fe^{2+}), it can react with $O_2^{\bullet -}$ to release $OH^{\bullet}$ (Free and Valentine 1977). This $OH^{\bullet}$, which is very aggressive for membranes, provokes lipid peroxidation, hyaluronic acid depolymerization, inactivation of essential enzymes and inhibition of some transport proteins (Rao et al. 1987). Similarly, another class of strong oxidant agents, such as hypohalogenic acids (HOX) exists. Some elements, such as chloride or iodide, can react with H_2O_2 in the presence of a peroxidase. For example, in the presence of chloride and myeloperoxidase, hypochloric acid is formed and reacts very rapidly with primary amines to give *N*-chloramines (Rao et al. 1987). Hypochloric acid and *N*-chloramines, which are known to be very lipophilic, are very cytotoxic. They usually increase sulphydryl group oxidation and essential enzyme inactivation; they can induce mutagenesis of nucleic acids and damage numerous amino acids and proteins.

A physiological balance exists between oxygenated free radical release and their catabolism by antioxidant agents. Some enzymatic systems, such as superoxide dismutase (SOD), catalase and glutathione peroxidase, are considered to be the first line of protection (Rao et al. 1987; McGahan and Fleischer 1986). SOD, which is a metalloprotein, catalyzes the rapid transformation of $O_2^{\bullet -}$ into H_2O_2 and molecular oxygen (O_2). This enzyme is present in outer segments of photoreceptors and at the level of pigment epithelium. Catalase, hemoprotein also present in rod outer segments and pigment epithelium, catalyzes the transformation of H_2O_2 into water and O_2. Glutathione peroxidase, present in the cytosol and the mitochondrial compartment, can prevent, in the presence of reduced glutathione, the accumulation of H_2O_2 in the intracellular medium, transforming it into water and oxidized glutathione.

The second line of defense of the retina is constituted by molecules such as vitamins E and C, which are known to scavenge free radicals. At this level, various exogenous molecules can obviously interfere to increase protection against oxidative mechanisms.

The deleterious effects of oxygenated radicals can affect in vivo various biomolecules: proteins, enzymes, polysaccharides, nucleic acids and polyunsaturated fatty acids (Armstrong et al. 1981). The reactivity of oxygenated radicals towards unsaturated lipids is particularly important because, on one hand, it

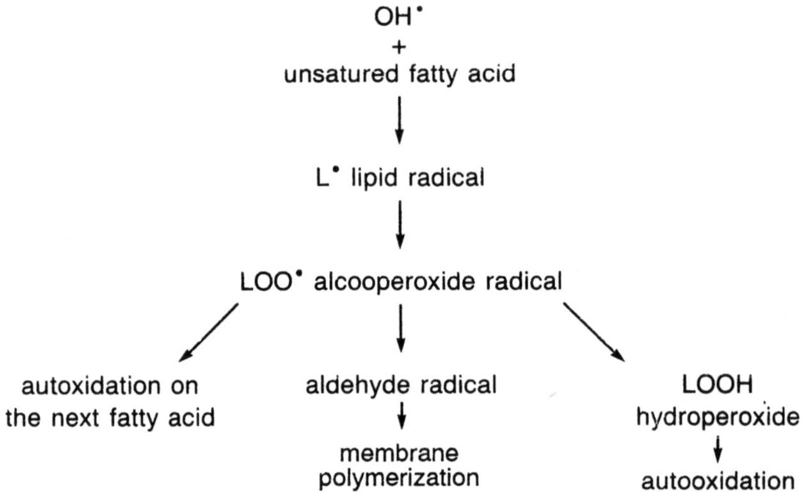

Fig. 1. Hydroxyl radical (OH•) reaction on unsaturated fatty acid and its consequences on membrane structure

induces the auto-oxidation process and, on the other hand, it destroys essential constituents of cellular membranes. The chemical attack of the lipid molecule provokes formation of a free lipid radical ($L^\bullet$) which, in the presence of oxygen, generates lipoperoxide radicals ($LO_2^\bullet$). These can react with many molecules, inducing an auto-oxidative process and membrane lysis (Fig. 1). At this level, protective agents may even act in blocking the auto-oxidative system and in protecting cellular membranes from deleterious effects of $LO_2^\bullet$ (Hayes 1974).

Oxygenated Free Radicals and Retinal Function

The retina is particularly rich in polyunsaturated fatty acids (Kagan et al. 1973), and thus represents an excellent model to test the biological effects of peroxide radicals and antioxidant molecules. Moreover, as mentioned previously, retinal function can be appreciated by recording the ERG, a quantitative electrophysiological signal whose generation implies the perfect integrity of cellular membranes (Hiramitsu et al. 1976; Shvedova et al. 1979). We report here our own results obtained on albino rat isolated retina (Doly et al. 1984, 1985).

After dark adaptation of the animal, the isolated retina, containing practically only rods, is maintained by perfusion and its functioning is ascertained by ERG recordings. The formation and release of free radicals ($OH^\bullet$) are induced by the addition, into the perfusion solution of a mixture of ferrous sulfate ($FeSO_4$) and sodium ascorbate. Fe^{2+}, maintained at valence II by sodium ascorbate, react with H_2O_2: $Fe^{2+} + H_2O_2 \rightarrow OH^\bullet + OH^- + Fe^{3+}$. $OH^\bullet$ can trap a hydrogen atom of an unsaturated fatty acid, thereby initiating the lipoperoxidation process. The functional consequences are rapidly significant; just after the peroxidative system addition, the ERG amplitude (b wave) decreases (Fig. 2).

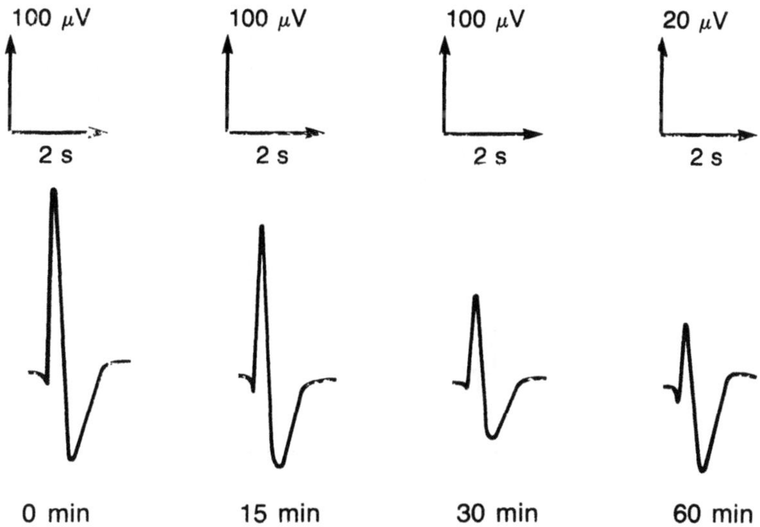

Fig. 2. Electroretinograms recorded on an isolated rat retina at various times after induction of retinal lipoperoxidation by adding to the perfusion solution the nonenzymatic peroxidative mixture: $FeSO_4$ (62.5 μM) + Na ascorbate (2.5 m*M*)

About 1 h after the initiation of lipoperoxidation, b wave ERG amplitude is no longer measurable, whereas in normal conditions, b wave remains measurable for 4-5 h after its stationary phase (Fig. 3). In this model it is possible to test the protective effect of various substances, such as free radical scavengers, which partially inhibit the deleterious consequences of free radicals on ERG and therefore, on retinal function. Figure 4 presents results obtained with SOD and figure 5 presents results obtained with a known free radical scavenger; they demonstrate partial protection of the ERG.

Comparable results can be obtained on rabbits when lipoperoxidation is induced by exposure to very high oxygen concentration (Hiramitsu et al. 1976). Over 2 days of observation, one can follow the progressive increase of $LO_2^{\bullet}$ concentration in tissues connected with a decrease of ERG amplitude. Histological examinations, performed at the end of the experiments, have established an unambiguous degeneration of photoreceptor cells. Likewise, intravitreous injection of $LO_2^{\bullet}$ induces an ERG impairment in the albino rabbit, with the signal nor being measurable after 7-10 days (Armstrong et al. 1982).

As previously mentioned, two membranes are implicated in the phototransduction mechanism which generates ERG: the plasma membrane and the saccular membrane of the photoreceptor. These two membranes are particularly rich in polyunsaturated fatty acids (Kagan et al. 1973; Oakley and Pinto 1981) and are therefore very sensitive to lipoperoxidation. The first one to be impaired and denatured by oxidation seems to be the saccular membrane. In fact, after lipoperoxidation, rhodopsin is more easily extracted (Novikov et al. 1975).

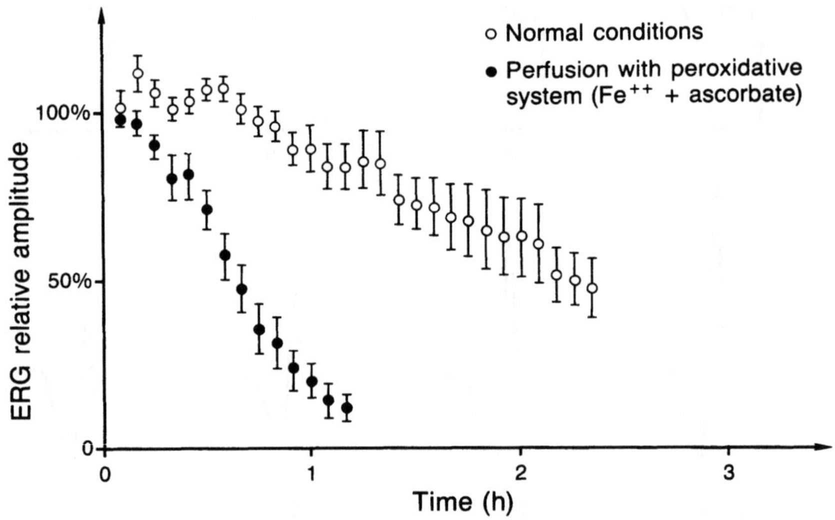

Fig. 3. Evolution of the amplitude of an electroretinogram (ERG) recorded on rat isolated retina in normal conditions (○) and in the presence of a peroxidative system: Fe SO_4 (62.5 μ*M*) + Na ascorbate (2.5 m*M*)

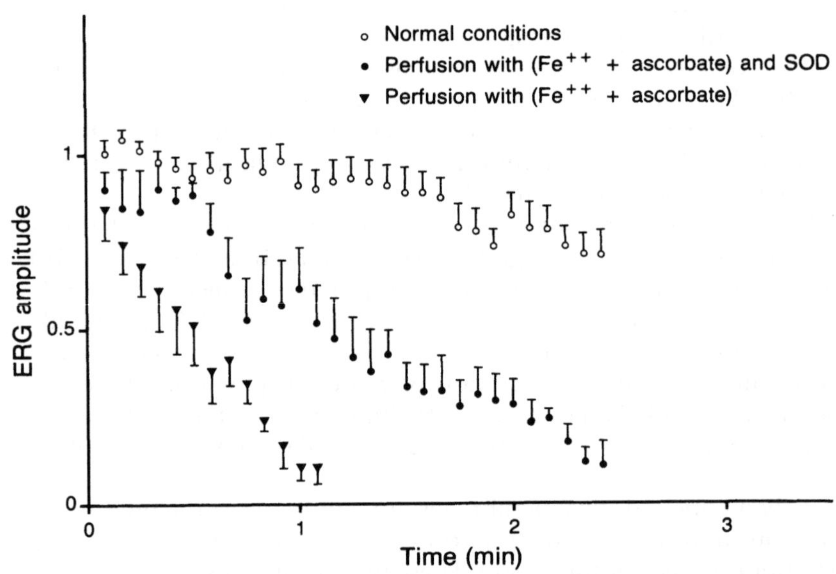

Fig. 4. Comparison of the evolutions of electroretinogram (ERG)) amplitudes recorded on isolated retina in the presence of the peroxidative system (Fe^{2+} + ascorbate) on untreated animals or on superoxide dismutase (SOD)-treated animals (15 000 U/kg SOD i. v., 45 min before isolation of the retina). ○ normal conditions; ● perfusion with peroxidative system on SOD-treated animals; ▾ perfusion with peroxidative system alone

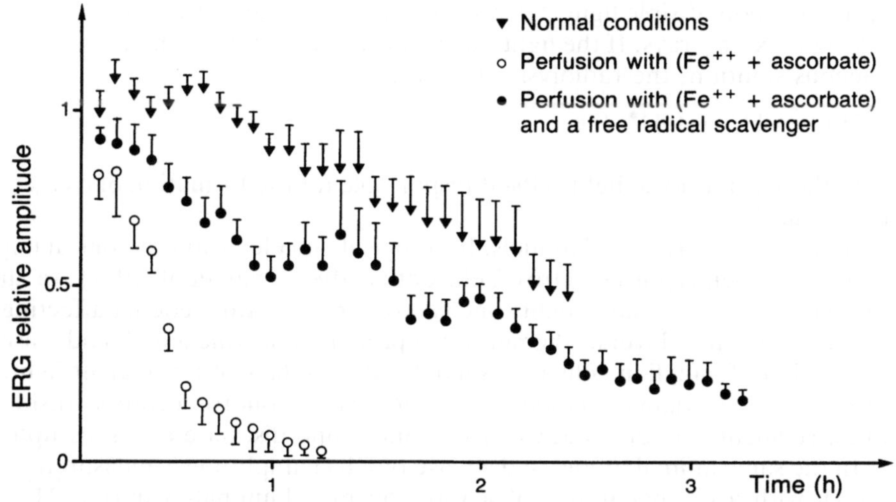

Fig. 5. Inhibition of the (Fe^{2+} + ascorbate)-induced lipoperoxidation effects on the electroretinogram (ERG) by perfusion in the presence of a free radical scavenger (Ginkgo biloba extract-EGB 761-100 mg/liter of perfusion solution). ▾ normal conditions; ○ perfusion with Fe^{2+} + ascorbate; ● perfusion with Fe^{2+} + ascorbate and a free radical scavenger

which may correspond to a weakening of protein-lipid interaction. Moreover, the regeneration capacity of rhodopsin decreases after peroxidation (Franswooth and Dratz 1976). Finally, intracellular recordings performed in rod outer segment indicate that lipoperoxidative substances interact essentially on the saccular membrane (Shvedova et al. 1979).

In the physiological state, the afore mentioned mechanisms of protection occur to preserve membrane integrity. However, in several pathological circumstances, these mechanisms may be insufficient, leading to specific impairments of retinal function.

Free Radicals and Retinopathies

In several retinopathies, oxygenated free radicals have been evoked. Free radical scavenging substances have been studied successfully in specific models. Retinal pathologies associated with overproduction of free radicals can be classed into two main groups.

Pathologies Resulting Directly from Free Radical Damage

Retinal photodegeneration

Retinal photodegeneration represents the most elementary impairment. Light which is the natural stimulating agent of the retina, is able, at high flux, to impair visual function. Visible light is part of the electromagnetic radiation spectrum as UV light, X-or γ-rays. If the light flux is sufficiently intense, one can observe in aqueous solution, the radiolysis of water:

$$H_2O \rightarrow H^+ + OH^\bullet + e^-$$

In these conditions, light is itself responsible for the formation of oxygenated free radicals.

In several experimental models, histological and electrophysiological impairments have ben reported (Droy-Lefaix et al. 1988; Penn et al. 1987). In albino rat, 24 h after continuous light, one can observe a retinal edema affecting the different cellular layers. Pigmentary epithelium is thickened and vascular congestion of ganglion cells is observed. After 4 days of permanent light, the most important damage is located at the level of photoreceptors. Inside the outer segment, the disc stack is disorganized and discs are often disrupted.

In the same animal model, a decrease of ERG amplitude is measured (about 80%), with a disappearance of a wave on rats illuminated during 24 h. By contrast, ERG amplitude remains normal in animals maintained in a photopic environment and then illuminated. In these animals, protective systems against lipoperoxidation are particularly developed, with vitamin E concentration and glutathione peroxidase activity being higher than in normals. So it seems that retinal sensitivity is decreased in light-adapted rats, which seem to be protected against photo-induced alterations.

Retinopathy of Prematurity

Oxygenated free radicals are closely implicated in retinopathy of prematurity (Bougle et al. 1982; Frank et al. 1978). Hyperoxygenation consequences have been extensively described in the albino rat (Penn et al. 1988). Newborn animals were maintained for 2 weeks in 60% O_2 and compared to a control group maintained in normal air. ERGs were recorded for 6 weeks; b wave amplitude was smaller in the oxygenated animals but no significant difference was noted in the a wave. Histological observations indicated that impairments were located at the inner layers of the retina, particularly in the Müller cell layer. These cells seemed to be the preferential target of oxygenated free radicals generated by hyperoxygenation, whereas photoreceptors or nervous cells of the retina seemed more resistant. In these conditions, it is normal to note a decrease of b wave and not of a wave. Such a localized impairment has major functional consequences, with the amplitude of the b wave varying with incident light intensity. More recently, a histological study (Reynaud et al. 1991), performed on a rat model of retinopathy of prematurity clearly demonstrated the link between hyperoxygenation and quality of retinal vasculature. Newborn rats, maintained for 17 days in 80% O_2, showed a substantial loss of retinal capillaries, and retinas exhibited nonvascularized peripheral areas. When transferred to room air, the blood vessels began to cover the nonvascularized areas. Treatment by a radical scavenger, started at birth, reduced the surface of the peripheral nonvascularized retina (Table 1).

Table 1. Mean area (mm^2) of nonvascularized peripheral retina observed on newborn rats maintained during 17 days in 80% oxygen. Animals were treated with either Ginkgo biloba extract (EGb 100 mg/kg/day orally) or the vehicle.

Retinal quadrant	Vehicle	EGB	Reduction (%)
Superior	0.74 ± 0.41	0.61 ± 0.36	18
Temporal	1.17 ± 0.64	0.70 ± 0.24	41
Nasal	0.62 ± 0.33	0.58 ± 0.43	7
Inferior	1.57 ± 0.65	0.82 ± 0.33	48
Total	4.10 ± 0.41	2.71 ± 0.41	34

Animals were treated with Ginkgo biloba extract (EGb, 100 mg/kg/day orally) or with the vehicle.

Retinal Ischemia

Numerous experiments have established that a prolonged ischemia causes irreversible tissue necrosis, particularly at the level of the retina (Hansen et al. 1989; Hayreh et al. 1990). Experimental retinal ischemia performed by venous or arterial occlusion induces the appearance of necrotic lesions at the origin of neovascularization (Gomolin 1989). In the retina, pathology associated with ischemia is mainly due to production of oxygenated free radicals during reperfusion sequence (McCord 1985). These oxygenated free radicals occur because, during ischemia, activity is decreased and the reactivity of free radicals normally produced during reperfusion is increased (Droy-Lefaix 1991). Moreover, free radicals are also produced in excess during reperfusion in several ways:

1. mitochondrial respiration;
2. xanthine oxidase/xanthine deshydrogenase system;
3. catabolism of arachidonic acid accumulated in tissues during ischemia;
4. leukocytes activated by leukotrienes, arachidonic acid or other immunomediators (platelet activating factor, PAF) implicated in inflammatory impairments due to ischemia.

In experimental models such as the albino rat, histological impairments of the retina have been studied after 90 min ischemia performed by occlusion of the central retinal artery followed by a 24 h reperfusion sequence (Szabo et al. 1991). Histological examinations, performed after the ischemia-reperfusion sequence, established the formation of an edema in the inner plexiform layer. This edema may be characterized by the thickness of the plexiform layer and the neutrophil migration. Many changes in the inner nuclear layer could also be observed; there were more pyknotic nuclei, vacuolated spaces and degenerative changes in the ganglion cells. These histopathologic changes were limited mostly to the inner part of the retina; only small alterations were seen in the photoreceptor layer.

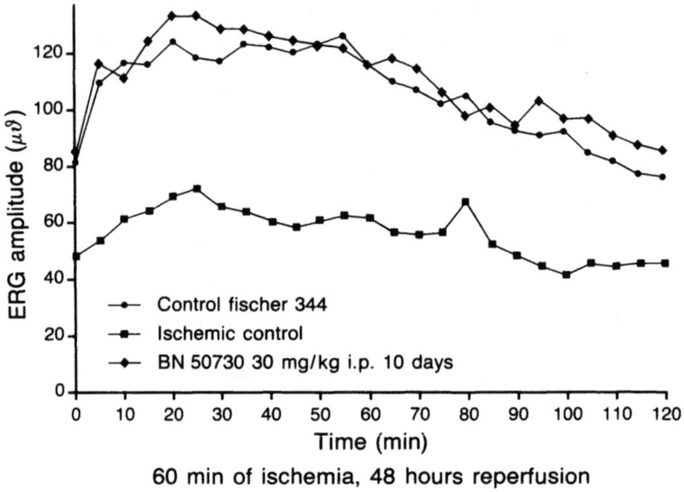

Fig. 6. Evolution of amplitudes, ERG recorded on isolated retina of: control Fischer 344 rats: (■); rats after 60 min ischemia followed by 48 h reperfusion (●); or treated (PAF antagonist BN 50730, 30 mg/kg i. p., 10 days) rats (◆) after 60 min ischemia and 48 h reperfusion

If ischemia duration is extended to 2 h and reperfusion to 48 h (Stefansson et al. 1988), the mitotic index is markedly increased. Under these conditions, retinal ischemia is able to induce a proliferative response which may be at the origin of diabetic retinopathy.

In similar experiments performed on albino rats, it was possible to measure the consequences of the ischemia-reperfusion sequence on the electrophysiology of the retina. A 60 min period of ischemia followed by 48 h of reperfusion induced a significant decrease of b wave ERG amplitude (Doly et al. 1991; Fig. 6). The implication of oxygenated free radicals in such impairments is confirmed by results obtained experimentally. On a rat ischemia-reperfusion model, 90-minute occlusion of the artery followed by 24 h reperfusion provoked retinal damage. Intravenous administration of SOD 5 min before the reperfusion significantly protected the retina. We noted the disappearance of pyknotic nuclei and degenerative changes of neurons with a severe decrease of neutrophil migration (Szabo et al. 1991). In the same experimental model, chronic treatment with a free radical scavenger, the Ginkgo biloba extract (EGb 761; Gardès-Albert et al. 1990) reduced edema formation in the outer plexiform layer and neutrophil migration (Szabo et al. 1991; Figs. 7 and 8). Finally, PAF was also able to stimulate free radical release by leukocytes. Effects of PAF were confirmed in retinal ischemia using a PAF-antagonist (BN 50730). Pretreatment with such an antagonist (BN 50730) induced practically a total recovery of ERG recorded on isolated retina albino rats after 60 min ischemia and 48 h reperfusion (Fig. 6).

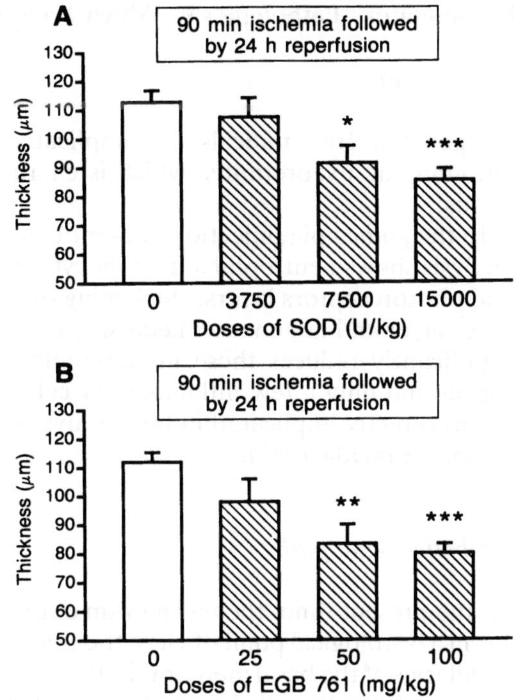

Fig. 7A, B Significant dose-related reduction of outer plexiform layer thickness after 90 min ischemia and 24 h reperfusion by *A* SOD administration (5 min before reperfusion) or *B* Ginkgo biloba extract (EGb 761), prolonged treatment. * $p<0.05$, ** $p<0.01$, *** $p<0.001$

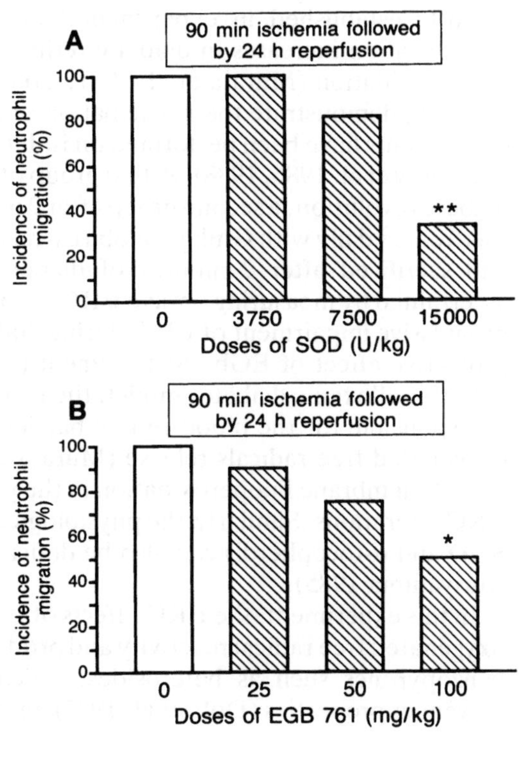

Fig. 8A, B Reduction of neutrophil migration observed in the retina after 90 min ischemia and 24 h reperfusion by *A* SOD administration (5 min before reperfusion) or *B* Gingko biloba extract (EGb 761), prolonged treatment. * $p<0.05$, ** $p<0.01$

Inflammatory Pathologies in Which Free Radicals Are Indirectly Implicated

Uveoretinitis

Oxygenated free radicals are implicated in the genesis and evolution of autoimmune uveoretinitis, which is characterized by serious vascular impairments.

In the guinea pig, injection of S-antigen induces release of $O_2^{\bullet -}$. Histological analysis shows neutrophil and monocyte infiltration into vitreous ganglion cells and photoreceptors layers, thickening of choroid, polynuclear accumulation in inner layer and large vessel necrosis (Yamada et al. 1986). SOD administration significantly reduces these impairments; choroid and vessels have a normal aspect and only a few inflammatory cells are seen in the retina. $O_2^{\bullet -}$ release seems directly implicated in human uveitis, resulting in membrane lipoperoxidation (Yamada 1983).

Diabetic Retinopathy

At the present time, the mechanism of diabetic retinopathy remains unknown. From a histological point of view, the characteristic impairment is a thickening of capillaries of the basal membrane (Bloodworth and Moditor 1985; Tanigushi and Nomura 1968; Papachristodoulou and Heath 1977), with a partial disruption of the blood-retinal barrier (Cunha-Vaz et al. 1975). However, several authors have recently established in experimental models that oxygenated free radicals concentration increases in diabetic retina with an enhancement of membrane lipoperoxidation (Murata et al. 1981; Cohen 1984; Nishimura and Kuriyamma 1985). To demonstrate the participation of free radicals in diabetic retinopathy, experiments have been performed on isolated retina of alloxan-induced diabetic rats (Doly et al. 1986, 1988). A first group of animals was sacrificed 1 month after diabetes evolution, a second group after 2 months of evolution, and a third group was treated daily with Ginkgo biloba extract (EGb 761), a free radical scavenger, and sacrificed after 2 months of diabetes evolution. Retinal function was ascertained by measuring b wave ERG amplitude. Our results clearly establish a progressive impairment of ERG during diabetes evolution (Fig. 9) and a partial protective effect of EGb 761 treatment (Fig. 10).

In this alloxanic diabetes model, the increase of capillary permeability, which is responsible for the blood-retinal barrier disruption, may be correlated with oxygenated free radicals release (Murata et al. 1981). Moreover, the elevated rate of membrane lipoperoxidation in the photoreceptors can be responsible for ERG alterations. Similarly, the physiological protective mechanisms (essentially SOD and α-tocopherol) can also be deficient or overwhelmed (Nishimura and Kuriyamma 1985).

In this experiment, the ERG effects of EGb 761 could be correlated both with oxygenated free radical reactivity and production. Effectively, EGb 761 is rich in benzopyrones such as heterosides, which are known for their free radical scavenger properties (Doly et al. 1985) and in Ginkgolides A, B and C (Braquet

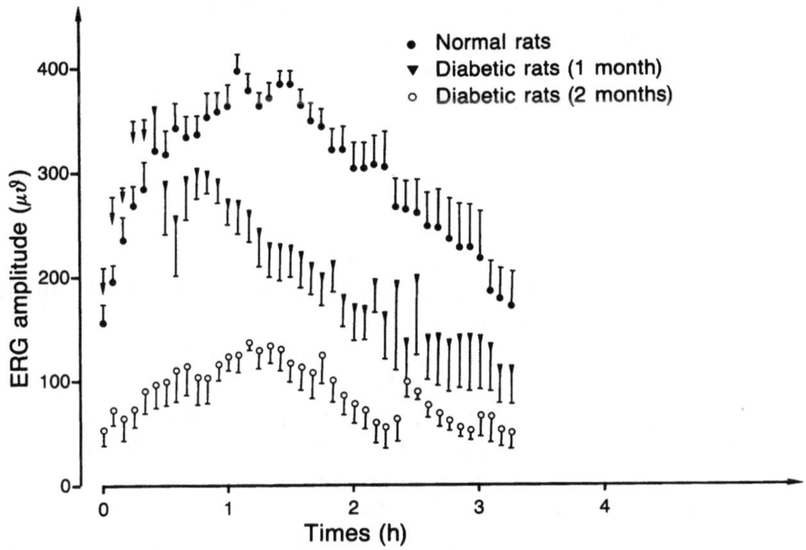

Fig. 9. Comparison of electroretinogram (ERG) amplitude evolution (isolated retina) between normal rats (●) and alloxan-induced diabetic rats after 1 (▼) or 2 (○) months' evolution of diabetes

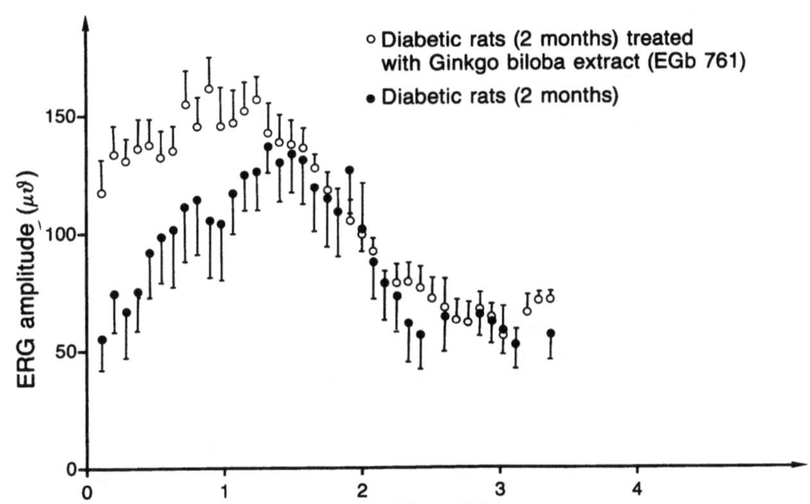

Fig. 10. Comparison of electroretinogram (ERG) evolution (isolated retina) on untreated (●) and Ginkgo biloba extract (EGb 761; 100 mg/kg/day) treated (○), aloxan-induced diabetic rats (2 months diabetes evolution)

et al. 1987), which are potent PAF-acether antagonists. PAF is an immunomediator able to amplify eosinophil and leukocyte responses and to induce release of $O_2^{\bullet -}$ and $OH^{\bullet}$ radicals. These radicals may be responsible for retinal membrane impairments and the dysfunctioning of diabetic retinopathy.

Toxic Retinopathy

Oxygenated free radicals are also implicated in several drug retinopathies. Recents experiments (Doly et al. 1990) demonstrated PAF and free radical participation in chloroquine retinopathy. Since 1950, chloroquine has been known for its retinal toxicity. Such a specific toxicity is active towards retinal neurons, photoreceptors and ganglion cells, but also towards retinal pigment epithelium (Raines et al. 1989). Generally, these impairments are reversible and are responsible for a decrease of visual acuity and alteration of ERG and electro-oculogram (Hodgkinson and Kobb 1970). To elucidate the mechanism of retinal alteration, we performed ERG recordings on albino rat isolated retina in the presence of 10^{-6} *M* chloroquine in the perfusion liquid. Two groups of animals were included in the experiment: an untreated group and a group treated with a specific PAF antagonist, BN 50730. Results indicated that chloroquine had a specific deleterious effect on ERG amplitude. By contrast, when chloroquine was added in the medium perfusing treated retinas, ERG amplitude was practically unaffected (Fig. 11). Such a remarkable protection confirms PAF implication in functional retinal impairments due to chloroquine. Oxygenated free radicals, released by leukocyte activation, are also probably directly implicated because recent results obtained on rats indicate a progressive

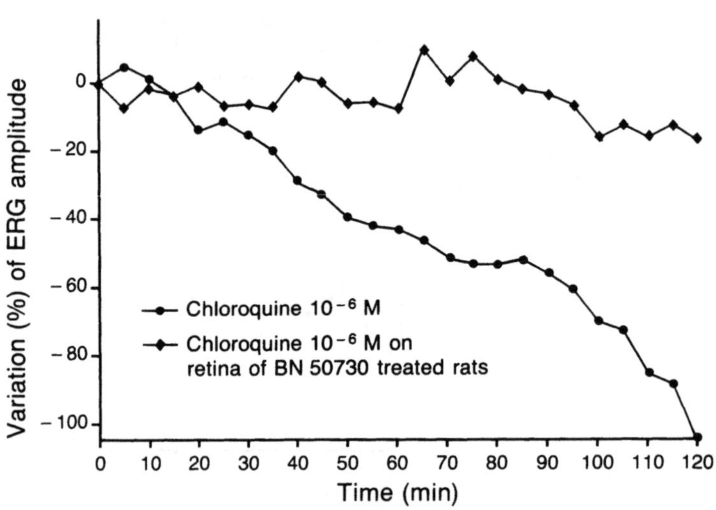

Fig. 11. Evolution of the variation (%) of electroretinogram (ERG) amplitude recorded in the presence of 10^{-6} *M* chloroquine on retina isolated from normal rats (●) or BN 50730 (PAF antagonist) treated (30 mg/kg/i.p., 5 days) rats (◆)

decrease of ERG amplitude in chloroquine-treated rats and a marked stabilization of ERG amplitude in rats treated with both chloroquine and a free radical scavenger (Doly et al., unpublished observations). This result confirms the significant participation of free radicals in the pathogenesis of chloroquine retinopathy and probably in other toxic retinopathies.

Conclusion

Oxygenated free radicals and $LO_2^{\bullet}$ that are produced and released into the retina play a key role in numerous retinopathies. In the retinal tissue the preferential targets of these radicals are the cellular membranes rich in polyunsaturated fatty acids. As these membranes are closely involved in the mechanism of phototransduction, oxygenated free radicals may be responsible for photoreceptor disorganization and retinal dysfunction.,

Drugs such as antioxidants, free radical scavengers or inflammatory reaction inhibitors may prevent or reduce functional impairments of various retinopathies. However, therapeutic use of these drugs must take into account some characteristic properties of free radicals. For instance, their half-lives are very limited and under these conditions, such anti-radical treatments are essentially preventive.

References

Armstrong D, Santangelo G, Connole E (1981) The distribution of peroxide regulating enzymes in the canine eye. Curr Eye Res 1:225-242

Armstrong D, Hiramitsu T, Gutteridge J, Nilsson SE (1982) Studies on experimentally induced retinal degeneration: 1. Effect of lipid peroxides on electroretinographic activity in the albino rabbit. Exp Eye Res 35:157-171

Bloodworth JHB, Moditor DL (1985) Ultrastructural aspect of human and canine diabetic retinopathy. Invest Ophthalmol Vis Sci 4:1037-1048

Bougle D, Vert P, Reichart E et al. (1982) Retinal superoxide dismutase activity in newborn kittens exposed to normobaric hyperoxia: effect of Vitamin E. Pediatr Res 16:400-402

Braquet P, Touqui L, Shen TY, Vargaftig BB (1987) Perspectives in platelet activating factor research. Pharmacol Rev 39:97-145

Cohen G (1984) Oxy-radical production in alloxan-induced diabetes: an example of an in vivo metal-catalyzed Haber-Weiss reaction. In: Armstrong D, Sohal RS, Cuther RG, Slater TF (eds) Free radicals in moecular biology. Raven, New York, pp 307-316

Cunha-Vaz J, De Abreu RF, Campos AJ, Figo GH (1975) Early breakdown of the blood-retinal barrier in diabetes. Br J Ophthalmol 59:649-656

Doly M, Braquet P, Bonhomme B, Meyniel G (1984) Effects of lipid peroxidation on the isolated rat retina. Ophthal Res 16:292-296

Doly M, Braquet P, Droy MT et al. (1985) Effects des radicaux libres oxygénés sur l'activité électrophysiologique de la rétine isolée de rat. J Fr Ophtalmol 3:273-277

Doly M, Droy-Lefaix MT, Bonhomme B, Braquet P (1986) Effets de l'extrait de Ginkgo biloba sur l'électrophysiologie de la rétine isolée de rat diabétique. Presse Med 15:1480-1483

Doly M, Braquet P, Droy MT et al. (1988) Alteration of electrophysiological function of isolated retina from alloxan-induced diabetic rats: effect of treatment with Ginkgo biloba extract. Neurochem Pathol 8:15-26

Doly M, Droy-Lefaix MT, Millerin M, Braquet P (1990) Implication of platelet-.activating factor (PAF) in retinal damage induced by ischemia-reperfusion. Invest Ophthalmol Vis Sci 32:1031

Doly M, Millerin M, Droy-Lefaix MT, Braquet P (1991) Inhibition des atteintes de l'électroretinogramme dues à la chloroquine par un nouvel antagoniste du PAF, le BN 50730. In: Christen Y, Doly M, Droy-Lefaix MT (eds) Physiologie, pathologie et génétique oculaires. Springer, Berlin Heidelberg New York, pp 155-160

Droy-Lefaix MT (1991) Ischémie rétinienne: pharmacologie et thérapeutique. In: Christen Y, Doly M, Droy-Lefaix MT (eds) Retinopathies et neurotransmission. Springer, Berlin Heidelberg New York, pp 63-74

Droy-Lefaix MT, Ruchoux MM, Guillemain J (1988) Protective effect of Ginkgo biloba extract (EGB 761) on light damage in the rat retina. J Cell Biochem 12:234

Fee JA, Valentine JS (1977) Chemical and physical properties of superoxide. In: Michelson AM, McCord JM, Fridovich I (eds) Superoxide and superoxide dismutase. Academic, New York, pp 19-57

Frank L, Bucher JR, Roberts RJ (1978) Oxygen toxicity in neonatal and adult animals of various species. J Appl Physiol 45:699-702

Fransworth CC, Dratz EA (1976) Oxidative damage of retinal rod outer segment membranes and the role of vitamin E. Biochim Biophys Acta 443:556-570

Freeman BA, Crapo JD (1983) Free radicals and tissue injury. Lab Invest 47:412-426

Fridovich I (1976) Oxygen radicals, hydrogen peroxide and oxygen toxicity. In: Pryor (ed) Free radicals in biology. Academic, New York, pp 239-279

Fridovich I (1983) Superoxide radical: an endogenous toxicant. Annu Rev Pharmacol Toxicol 23:239-279

Gardès-Albert M, Sekaki A, Ferradini C, Droy-Lefaix MT (1990) EGB 761 scavenger effect against $OH^{\bullet}$ and $O_2^{\bullet -}$ free radicals. A radiolysis study. Free Radic Biol Med 9:190

Gomolin JES (1989) Efficacy of panretinal photocoagulation in central retinal vein occlusion. Ophthalmologica 199:24-27

Hansen LL, Wiek J, Schade M, Muller-Stolzenburg N, Wiederholt M (1989) Effect and compatibility of isovolaemic haemodilution in the treatment of ischaemic and non-ischaemic central retinal vein occlusion. Ophtalmologice 199:90-99

Hayes KC (1974) Retinal degeneration in monkeys induced by deficiencies of vitamin E or A. Invest Ophthalmol Vis Sci 13:449

Hayreh SS, Klugman MR, Podhajsky P, Servais GE, Perkins ES (1990) Argon laser panretinal photocoagulation in ischemic central retinal vein occlusion. Graefes Arch Clin Exp Ophthalmol 28:281-296

Hiramitsu T, Hasegawa Y, Hirata K et al. (1976) Formation of lipoperoxide in the retina of rabbit exposed to high concentration of oxygen. Experientia 32:622-623

Hodgkinson BJ, Kobb H (1970) A preliminary study of the effect of chloroquine on the rat retina. Arch Ophthalmol 84:509-515

Kagan VE, Shvedova AA, Novikov KN, Koslov YP (1973) Lipid-induced free radical oxidation of membrane lipids in photoreceptors of frog retina. Biochim Biophys Acta 330:76-79

McCord JM (1985) Oxygen-derived free radicals in post-ischemic tissue injury. N Engl J Med 312:159-163

McGahan MC, Fleischer LN (1986) Antioxidant activity of aqueous and vitreous humor from the inflamed rabbit eye. Curr Eye Res 5:641-645

Mittag T (1984) Role of oxygen radicals in ocular inflammation and cellular damage. Exp Eye Res 39:759-769

Murata R, Nishida T, Eto S, Mukai N (1981) Lipid peroxidation in diabetic rat retina. Metab Pediatr Syst Ophthalmol 5:83-87

Nishimura C, Kuriyamma K (1985) Alteration of lipid peroxide and endogenous antioxidant contents in retina of streptozotocin-induced diabetic rats: effect of vitamin A administration. Jpn J Pharmacol 37:365-372

Novikov KN, Kagan Ve, Shvedova AA, Kozlov YP (1975) Protein-lipid interactions on peroxide oxidation of lipids in the photoreceptor membrane. Biofizika 20:1039-1042

Oakley B. Pinto LH (1981) $[Ca^{2+}]_i$ modulations of membrane sodium conductance in rod outer segments. Curr Top Membr Transp 15:405

Papachristodoulou D, Heath H (1977) Ultrastructural alterations during the development of retinopathy in sucrose-fed and streptozotocin-diabetic rats. Exp Eye Res 25:371-384

Penn JS, Naash MI, Anderson RE (1987) Effect of light history on retinal antioxidants and light damage susceptibility in the rat. Exp Eye Res 44:779-788

Penn JS, Thum LA, Rhem MN, Dell SJ (1988) Effects of oxygen rearing on the electroretinogram and GFA-protein in the rat. Invest Ophthalmol Vis Sci 29:1623-1630

Raines MF, Bhargava SK, Rosen ES (1989) The blood-retinal barrier in chloroquine retinopathy. Invest Ophthalmol Vis Sci 30:1726-1731

Rao NA, Romero JL, Fernandez MA et al. (1987) Role of free radicals in uveitis. Surv Ophthalmol 32:209-213

Reynaud X, Vallat M, Vincent D, Dorey CK (1991) Protective effect of the Ginkgo biloba extract in the rat model of retinopathy of prematurity (ROP). Invest Ophthalmol Vis Sci 32:1147

Shvedova AA, Sidorov AS, Novikov KN et al. (1979) Lipid peroxidation and electric activity of the retina. Vision Res 19:49-55

Stefansson E, Wilson Ca, Schoen T, Kuwabara T (1988) Experimental ischemia induces cell mitosis in the adult rat retina. Invest Ophthalmol Vis Sci 29:1050-1055

Szabo ME, Droy-Lefaix MT, Doly M, Carré C, Braquet P (1991) Ischemia and reperfusion-induced histologic changes in the rat retina. Invest Ophthalmol Vis Sci 32: 1471-1478

Tanigushi Y, Nomura T (1968) Fine structure of retinal blood vessels in human diabetics. Acta Soc Ophthalmol Jpn 72:1165-1178

Yamada M (1983) Lysosomal enzyme activities and superoxide generation of neutrophils in patients with Behcet disease. Folia Ophthalmol Jpn 34:1678-1685

Yamada M, Shichi H, Yuasa T et al. (1986) Superoxide in ocular inflammation: human and experimental uveitis. Free Radic Biol Med 2:111-117

Free Radical Activity in Chronic Ethanol Intoxication in Humans

T. L. Dormandy

Introduction

Chronic ethanol poisoning is probably the commonest chemical intoxication in the world and yet one of the least understood. It can also be one of the most difficult diseases to diagnose in clinical practice. Much indirect evidence based on cell, tissue and animal studies suggests – and has done so for many years – that it is associated with increased, or at least altered, free radical activity. In humans, unfortunately, the evidence is much more sparse and far more difficult to interpret.

Possible Sources of Free Radicals

Theoretically one can envisage at least five mechanisms which could lead to increased free radical generation in ethanol intoxication. First, ethanol itself could generate the ethoxy radical which in turn could initiate other free radical reactions. There is some laboratory electron spin resonance (ESR) evidence for this (Albano et al. 1988). Second, chronic ethanol intoxication is associated with increased cytochrome P450 activity – tenfold or more – in the liver (Ekstrom et al. 1988; Thurmann et al. 1972). This and perhaps other detoxification mechanisms could be an important source of escaping free radicals. Third, it has been suggested that free radicals might be generated in areas of hypoxia caused by the accumulation of acetaldehyde, a possible reperfusion-injury scenario (Lewis and Paton 1982). Fourth, infiltration by inflammatory cells is an invariable feature of chronic ethanol toxicity in many tissues at some stage of the disease, and these cells could generate reactive oxygen species and lead to lipid peroxidation (Shaw et al. 1981). Fifth, one has to recall that structural cell damage of virtually any kind will facilitate free radical generation (Dormandy 1986).

In addition to these different possible mechanisms of free radical generation one must also distinguish between at least three forms of ethanol toxicity.

First, there is acute ethanol intoxication in a previously fit, i.e., nonalcoholic, individual. This is almost entirely a neurological syndrome; even the often prominent gastrointestinal manifestations are probably attributable to autonomic neuropathy rather than to a direct effect of alcohol on the gastrointestinal tract. (I would mention, incidentally, that the same is true of acute carbon tetrachloride poisoning.) The condition can be fatal; about 500 deaths may be caused by it every year in the UK.

Second, there is progressive alcoholic liver disease, passing through a number of overlapping stages - acute inflammatory, degenerative and chronic inflammatory. One has to assume that the disease exists in every individual who has consumed over 100 g ethanol a day for a year or more, irrespective of the prominence or social acceptability of his or her symptoms. Other organs and systems, e.g., the pancreas, the parotid, the haemopoeitic and nervous system, are affected to a variable degree.

Third, there are a number of atypical forms of chronic ethanol intoxication in which certain tissues or organs are affected out of proportion to others. A relatively mild skeletal myopathy, for example, is common in chronic alcoholics, but occasionally life-threatening muscle paralysis dominates the clinical picture. In the same way a neuropsychiatric syndrome may quite eclipse all other manifestations.

Acute Ethanol Intoxication

This is the only form of ethanol intoxication that can easily be produced, though not necessarily reproduced, in animals. In mice, rats and rabbits some workers have reported an increase in thiobarbituric acid (TBA) reactive material in the liver and some other organs. However, since there are striking enzyme inhibitions to account for the physical signs and death, this increase has never been pursued with much vigour. Information about acute intoxication in previously normal, i.e., not chronically addicted, men or women is virtually nonexistent. You may or may not agree that this is a sad reflection of the lack of liaison between groups of people interested in alcohol research but determinedly ploughing their own furrow. Over a 5-year period, when some of my colleagues were investigating the problem they were able to obtain material suitable for biochemical studies in only three cases. Ours, therefore, is anecdotal evidence. We did not find an increase in (TBA)-reactive material – the only free radical marker available at the time – in any of these cases, but the brain, liver and several other organs we examined showed a dramatically increased susceptibility to peroxidation when exposed to oxidant stress.

Chronic Ethanol Toxicity

Biochemical research into chronic ethanol toxicity in humans has to contend with two difficulties. First, except possibly in baboons, the disease cannot be credibly reproduced in animals. Second, as in most chronic diseases, the pathology of the disease reflects a combination of two mechanisms, that of the causative process (in this case repeated chemical insults) and that of the body's defence response (in this case detoxification mechanisms). It is often extremely difficult to distinguish between the two, all the more so since in vitro and in vivo evidence may contradict each other. In most experimental systems, for example, methanol and ethanol (like ethanol and carbon tetrachloride) act synergistically, mutually aggravating each other's toxic effects. Yet in clinical practice there is no doubt that ethanol addicts can consume many times the fatal amount of methanol (or carbon tetrachloride) with comparative impunity. Nevertheless the condition has been studied with two more or less overlapping objectives in mind. First, there is a pressing need for better diagnosis. (Let us recall that it is one of the few conditions in which the history, usually the mainstay of any clinical diagnosis, is often positively misleading.) Second, it is reasonable to assume that a better understanding of possible free radical involvement in the underlying pathology might contribute to more effective treatment.

Changes in Blood

Inevitably all "diagnostic" studies have been based on plasma and red blood cells, the only tissue or tissue fluid readily available in routine clinical practice. There have been a number of relatively promising new laboratory methods proposed in recent years – the measurement of aldehyde-protein adducts or of desialated transferrin to mention but two (Schellenberg and Weill 1987; Carini et al. 1988) – but the most widely used combination of tests, the serum γ-glutamyltransferase activity and the mean red cell corpuscular volume, are positively diagnostic in only about 70% of cases. Reports of increased TBA-reactive material in plasma, decreased plasma thiols, a depressed plasma antioxidant potential and diminished red cell superoxide dismutase activity have not been very convincing (at least in a diagnostic context).

The measurement of a diene-conjugated derivative of linoleic acid in relation to the nondiene-conjugated parent compound, usually expressed as the ratio 18:2(9, 11) × 100/18:2(9, 12) in plasma phospholipids is still, to my mind, one of the most useful markers (Iversen et al. 1985; Cawood et al. 1984). I say "still" because there have been several developments since the finding was originally described. I would mention three in particular (Fink et al. 1985; Dormandy 1988).

First, contrary to my own prognostications a few years ago, it is now clear that the plasma concentration of the diene-conjugated lipid 18:2 (9,11) is strongly influenced by the long-term diet. In particular, it is increased by a high level of consumption of milk and dairy products and, conversely, extremely low in populations (such as some Pacific Islanders) who are genetically lactase-deficient and consume virtually no milk, However, this dietary effect is a long-term one; consuming half a pint of cream on the one hand or starvation for 24 h on the other has virtually no effect. Many chronic alcoholics do of course keep to a bizarre diet, but it is almost always low in milk products.

Second, microorganisms which can convert linoleic acid, (18:2(9,12), to a succession of diene-conjugated derivatives, (18:2(9,11), have been shown in the last few years to be common nonpathogenic commensals on the human skin and mucous membranes and in secretions such as saliva and cervical mucous

(Fairbank et al. 1988). The action of these organisms is similar to that of organisms which flourish in the rumen of cattle and whose activity accounts for the high concentration of diene-conjugated linoleic acid isomers in the plasma and milk of ruminants. The normal human stomach and small intestine are, of course, for practical purposes sterile, and diene-conjugating activity is destroyed by gastric acidity, but neither of these conditions necessarily applies to chronic alcoholics.

Third, we now realise that while "free", i.e., non-esterified, fatty acids can be virtually ignored in lipid analyses of normal plasma and in the vast majority of diseases, in certain conditions they can rise from less than 1 % to as much as 40 % of the total. These include most strikingly the lipolytic state induced by heparin and diabetic coma, but the unesterified fatty acid concentration can also be abnormally high in some alcoholics (Hind et al. 1988). This matters because, for reasons which are still not quite clear, as much as half of the unesterified fatty acids in plasma may be diene-conjugated (in contrast to 1 %–4 % of the esterified forms.)

With these provisos in mind, at least four well-conducted series on chronic alcoholics in different detoxification centres have shown that the 18:2 (9, 11)/18:2(9, 12) ratio in plasma phospholipids is significantly raised in at least 80 % of patients admitted and that it drops sharply and rapidly over the first few days of alcohol withdrawal. No such increase was seen in patients with acute hepatitis, patients on prolonged drug treatment (e.g., epileptics) or in normal pregnancy, but abnormal ratios were found in about half of our patients suffering from the rather mysterious condition of primary non-alcoholic biliary cirrhosis, in three patients with paraquat poisoning studied in another centre and in some women with pre-eclamptic toxaemia of pregnancy (Fink et al. 1985; Situnayake et al. 1990). However, the most characteristic feature in chronic alcoholics is their response to even small doses of alcohol. This is a rather controversial topic, since the administration (either intravenously or by mouth) of even a trace amount of ethanol to patients on withdrawal treatment may be regarded as unethical. However, in three patients who did procure an alcoholic drink for themselves during the first 2 weeks of withdrawal therapy, their relatively modest relapse produced a quite dramatic and sharp increase in their 18:2(9,11)/18:2(9,12) ratio. Normal subjects, by contrast, can consume half a litre of gin or whiskey, or perhaps more, without any change.

The overall pattern suggests that, in chronic alcoholics, a mechanism is induced which leads to the detoxification of ethanol by a mechanism that entails the conversion of linoleic acid to its diene-conjugated derivative and that this mechanism is activated by even a small load of ethanol. The question is of course how far this is a free radical-mediated rather than an enzymatic reaction. My own opinion is that it is both. Even in the microorganisms of ruminants the so-called enzymatic conversion of nondiene-conjugated to diene-conjugated fatty acids is probably a two-step reaction. The first step is the generation (probably by another free radical) of a fatty acid free radical, and the second step is a more specific enzymatic reduction of this free radical to a stable diene-conjugated isomer.

Liver Biopsies

Theoretically, much more information should be obtainable from studying tissues known to be affected in chronic ethanol intoxication, notably from liver biopsies. In fact these studies have been slightly disappointing mainly, I think, because the material is rarely sufficient for dynamic studies. This is of course a common limitation in biochemical work on human tissues, but it is especially restrictive in intoxications and detoxification mechanisms. Chemical analysis of a tissue at the moment of a biopsy is rarely a real advance on microscopic and ultramicroscopic techniques. What would be – or might be – a fundamental advance would be to study the response of these tissues to exposure to ethanol (or to other relevant toxic agents) in vitro.

However, chemical analysis of biopsy samples has not been without interest. The most recent study, by investigators at the Dudley Road Hospital, Birmingham, confirms earlier findings of a high level of lipid diene conjugation in alcoholic livers (Situnayake et al. 1990). This reflects a not very dramatic increase in lipid peroxides but a significant increase in the concentration of nonperoxide diene-conjugated lipids. I am fairly certain that earlier reports by American workers on chemical evidence of increased peroxidation in alcoholic livers have been based on misinterpreting increased diene conjugation, diene conjugation is in fact increased but reflects mainly nonperoxide isomers (Lieber 1984). In the Birmingham study the concentration of nonperoxide diene-conjugated isomers (but not of peroxides) correlated inversely and significantly with a depletion in reduced glutathione. It also correlated with histological appearances of inflammation and iron staining. The ratio of vitamin E to lipids was variable and not clearly related to other chemical or to histological changes.

For the reasons I mentioned it is not easy to interpret these findings. I do not think that the lipid changes reflect an abnormality in the diet or in intestinal absorption, though both may be contributing or complicating factors. The changes do, I think, reflect a significant increase in free radical activity but not necessarily in lipid peroxidation. I also think – but this is hardly more than guesswork – that the increase is due to an increase in detoxifying activity rather than to an increased generation of ethoxy radicals.

Other Free Radical Aspects of Ethanol Toxicity

It has been impossible in a brief review to discuss more than a few free radical aspects of a complex and varied condition. I would simply mention that severe ethanol-induced skeletal muscle paralysis is almost certainly related to an increased accumulation of TBA-reactive material and other evidence of free radical damage (Martin et al. 1985). Some evidence suggests that the characteristic macrocytosis of chronic alcoholics can be induced by incubating normal red cells in plasma rich in peroxidised lipoproteins. And finally, possible free radical damage to amino acids, especially to tryptophane, may be a factor in their diminished availability to the brain, perhaps an important mechanism in producing the neuropsychiatric manifestations of ethanol.

References

Albano E, Tomasi A, Goria-Gatti L, Poli G, Dianzani MU (1988) Spin trapping of ethanol derived free radicals in rat liver microsomes. In: Nordmann R, Ribière C, Rouache H (eds) Alcohol toxicity and free radical mechanisms. Pergamon, Oxford, pp 17–23

Carini R, Mazzanti R, Biasi F, Chiarpotto E, Marmo G, Moscarella S, Gentilini P, Dianzani MU, Poli G (1988) Fluorescent aldehyde-protein adducts in the blod serum of healthy alcoholics. In: Nordmann R, Ribière C, Rouache H (eds)Alcohol toxicity and free radical mechanisms. Pergamon Oxford, pp 61–69

Cawood P, Wickens DG, Iversen SA, Braganza JH, Dormandy TL (1984) The nature of diene conjugation in human serum. FEBS Lett 171: 320–324

Dormandy TL (1986) Free radicals in medicine. In: Triger DR (ed) Advanced medicine. Bailliere Tindall, London, pp 333–340

Dormandy TL (1988) Diene conjugation in chronic alcoholics. In: Nordmann R, Ribière C, Rouache H (eds) Alcohol toxicity and free radical mechanisms. Pergamon, Oxford, pp 55–60

Ekstrom G, von Bahr C, Ingelman-Sundberg K (1988) Ethanol-inducible cytochrome P-450 from human liver. In: Nordmann R, Ribière C, Rouache H (eds) Alcohol toxicity and free radical mechanisms, Pergamon. Oxford, pp 219–224

Fairbank J, Ridgway L, Griffin J, Wickens DG, Singer A, Dormandy TL (1988) Diagnosis of cervical neoplasia by the estimation of octadeca-9,11-dienoic acid. Lancet i: 537–539

Fink R, Clemens MR, Marjot DH, Iversen SA, Cawood P, Dormandy TL (1985) Increased free-radical activity in alcoholics. Lancet ii: 291–294

Hind CKR, Griffin J, Pack S, Latchman YE, Drake HF, Jones HM, Brostoff J, Dormandy TL, Treasure T (1988) The effect of cardiopulmonary bypass on circulating levels of free-radical products.Cardiovasc Res 22: 37–44

Iversen SA, Cawood P, Dormandy TL (1985) A method for the estimation of a diene-conjugated derivative of linoleic acid in serum phospholipids and its possible origin. Ann Clin Biochem 22: 137–140

Lewis K, Paton A (1982) Could superoxide cause cirrhosis? Lancet ii: 188–189

Lieber CS (1984) Alcohol and the liver. Hepatol Update 4: 1243–1260

Martin F, Ward K, Slavin G, Levi J, Peters TJ (1985) Alcoholic skeletal myopathy. Q J Med 55: 231–251

Schellenberg F, Weill J (1987) Desialotransferrin measured by isoelectric focusing. Ann Biol Chem 45: 290–295

Shaw S, Jayatilleke R, Ross WA, Gordon ER, Lieber CS (1981) Ethanol induced lipid peroxidation. J Lab Clin Med 98: 417–425

Situnayake RD, Crump BJ, Thurnham DI, Davies JA, Gearty J, Davis MN (1990) Lipid peroxidation and hepatic antioxidants in alcoholic liver disease. Gut 31: 1311–1317

Thurmann RG, Ley HC, Scholz R (1972) Hepatic microsomal ethanol oxidation. Eur J Biochem 25: 220–230

The Possible Role of Lipid Peroxidation in the Pathophysiology of Mental Disorders

L. Prilipko

Summary

It has been recognized that, in some circumstances, brain tissues can be damaged by oxidative stress. The data presented here demonstrate that stress in animals, as well as strong emotional strain in healthy men, can initiate lipid peroxidation (LPO).

Experiments on rats have shown that activation of lipid peroxidation in the brain due to stress can be the key factor in the damage of a number of membrane-bound structures (receptors, enzymes) of the neurons from the cerebral cortex. It has also been demonstrated that the level of lipid peroxidative products significantly increased in the serum of patients with manic depressive disorders and schizophrenia. The author considers the possibility of applying antioxidants as an important component in the combined treatment of some mental disorders, as well as for the prevention of stress-induced brain damage.

Introduction

In the last decade, specialists involved in biological research in psychiatry have shown a remarkable interest in investigations of the structure and function of biomembranes and membrance-bound proteins, such as receptors, enzymes, and ionophores. Studies along these lines have turned out to be quite successful. Thus it has been shown that, in many bipolar and some schizophreniform disorders, the phospholipid content in the membranes is significantly changed (Hitzemann and Garver 1981; Henn and Henn 1981). Schizophrenia has been found to be concomitant with changes in some neuronal receptors (Bennett et al. 1979; Zeeberg et al. 1990) and membrane enzymes (Wise et al. 1974; Reddy et al. 1990). But the cause of such kinds of changes in the membrane properties in patients with mental disorders remains unclear. Therefore, it seems worthwhile to investigate the mechanisms underlying modification and damage of membrane structures in nerve cells.

As a working hypothesis, it has been suggested that free radical peroxidation of polyene phospholipids can play a leading role in membrane modifications during the course of some mental disorders, as well as during strong emotional stress.

Until now, there have been very few reports in the literature concerning the possible role of LPO in the pathogenesis of mental disorders (Lohr et al., 1988). Brain tissue seems to be the most plausible target for oxidative stress, since the brain consumes the bulk of molecular oxygen utilized by the organism. Besides, brain tissues contain large amounts of phosopholipids, including unsaturated fatty acids, the main substrates of LPO, as well as a wide variety of enzymes responsible for the oxidation of biological substrates mediated by O_2 and its activated species capable of inducing LPO. Therefore, in the present work we are trying to investigate the role of LPO in stress-induced damage of brain membranes and to assess the level of this process in stress-exposed individuals and in manic-depressive patients as well as in schizophrenics.

Methodology and Results

The stress in test animals was reproduced by the procedure of Desiderato and Testa (1976). The intensity of LPO was studied by three different methods allowing a quantification of primary and secondary products of LPO accumulated in brain tissues in vivo. These procedures included a spectrophotometric assay, polarographic measurements with a drop of mercury electrode and determination of end LPO products by fluorescence of Schiff bases (Kagan et al. 1979).

The data listed in Table 1 demonstrate that exposure to stress resulted in an accumulation of primary and secondary products of LPO in all the tissues under investigation. It is important to note that, after a preliminary interperitoneal injection of the synthetic free radical scavenger 4-methyl-2.6-diterbutylphenol (BHT; 120 mg/kg of body weight daily for 3 days), the stress-induced activation

Table 1. Accumulation of lipid peroxidative products in rat tissues after the stress and protective effect of BHT

Tissue	Experimental group	Diene conjugates (o.d. unit/mg lipid)		Hydroperoxides (nmol/mg lipid)		Fluorescent Schiff bases (arbitrary units)	
		NaCl	BHT (120 mg/kg)	NaCl	BHT (120 mg/kg)	NaCl	BHT (120 mg/kg)
Brain	Control	0.15±0.02	0.15±0.02	0.5±0.1	0.4±0.1	7.8±0.9	10.2±0.7
	Stress	0.45±0.05	0.25±0.04	1.9±0.3[a]	0.5±0.1	20.7±3.2[a]	14.2±2.3
Retina	Control	0.51±0.06	0.50±0.02	–	–	4.7±0.3	4.2±0.4
	Stress	0.83±0.07[a]	0.52±0.03	–	–	11.4±0.7	5.2±0.5
Skeletal Muscle	Control	0.40±0.05	0.35±0.05	0.6±0.1	0.6±0.1	10.3±0.9	10.4±2.1
	Stress	0.80±0.05[a]	0.40±0.05	1.4±0.1[a]	0.8±0.2	20.6±2.8[a]	15.6±3.6
Myo-cardium	Control	0.35±0.03	0.35±0.01	1.3±0.4	1.2±0.3	10.5±3.5	10.6±1.2
	Stress	1.00±0.05[b]	0.60±0.05[a]	8.0±1.6[a]	1.5±0.3	29.2±4.6[a]	13.2±1.8

[a] $p<0.01$; [b] $p<0.001$

Table 2. Dynamics of content of the fluorescent Schiff bases in rat tissues due to stress

		Time elapsed after stress exposure			
Tissue	Control	2 h	2 days	5 days	10 days
Brain	7.8±0.9	20.7±3.2[a]	26.9±3.3[b]	23.7±4.1[a]	10.6±2.2
Skeletal muscle	10.3±0.9	20.6±2.8[a]	27.0±9.7[a]	10.8±1.7	9.0±1.2
Myocardium	10.5±3.5	29.2±4.6[a]	27.6±4.3[a]	12.9±1.9	9.0±1.1
Aorta	11.5±1.9	31.2±7.5[b]	23.0±2.8[a]	10.5±2.6	–

[a] $p<0.01$; [b] $p<0.001$

of LPO was completely abolished. An analysis of the dynamics of LPO products accumulation in rat tissues revealed that the LPO level in the brain markedly exceeded that in other tissues.

As can be seen from Table 2, on the fifth day after the stress exposure the intensity of LPO in all the tissues under investigation dropped to the initial level, while in the brain this decrease was observed only on the tenth day.

In the next experimental series, we studied the effects of LPO on the permeability of the hematoencephalic barrier. While doing this, we proceeded from the assumption that the brain is a beyond-the-barrier-organ, which implies that the appearance of brain-specific proteins in peripheral blood is immediately followed by the production of antibodies against brain antigens. In our study, this might be due to the changes in the permeability of the hematoencephalic barrier. The data shown in Fig. 1 suggest that, on the eighth day after the stress exposure, antibodies against water-soluble (column 1A) and membrane-bound antigens (column 1B) began to appear in the blood of the test animals. The level of brain-specific antibodies in the blood of stress-exposed animals was not increased if the BHT injections were done 3 days before the stress (columns 2 A and 2 B). BHT itself normally has no effect on antibody synthesis (columns 3 C and 4 C).

The changes in the properties of membrane-bound proteins can be illustrated by the results from the specific binding of [^{3}H] dihydroalprenalol by brain cortex membranes before and after the stress.

Table 3. Effect of stress-induced lipid peroxidation on specific binding of [^{3}H]dihydroalprenolol in the rat brain cortex

Influence	Fluorescence of Schiff bases (%)	Binding of [^{3}H]dihydroalprenolol (%)
Control (n=19)	100	100
BHT, 120 mg/kg (n=13)	105±13	96±12
Stress (n=11)	354±43	55±14
Stress + BHT (120 mg/kg) (n=11)	164±30	92±16

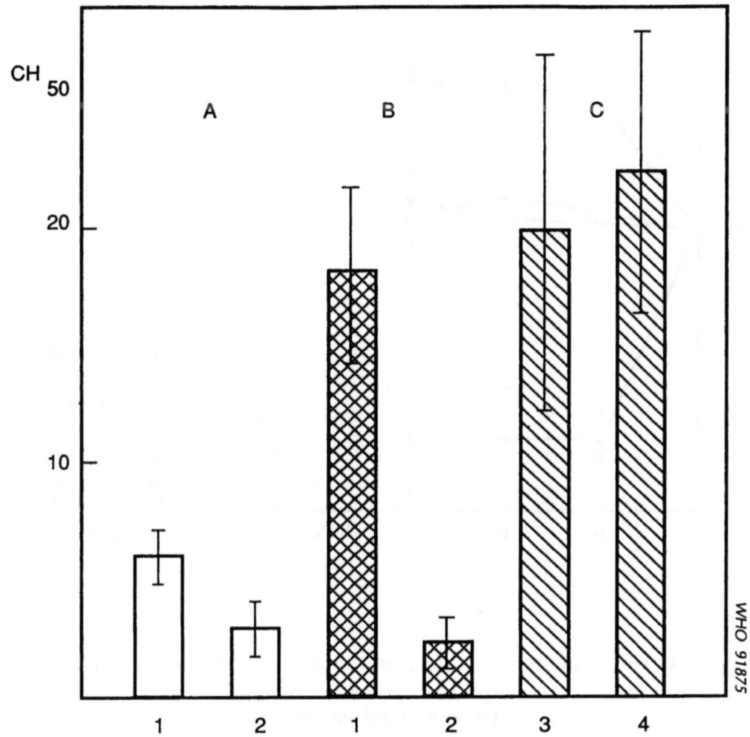

Fig. 1. Antibrain antibody production in Wistar rats exposed to stress. Ordinate, number of bound units of the complement CH_{50} (Bogdanova et al. 1981). A, during interaction with water-soluble brain antigen; B, during interaction with membrane-bound brain antigen; C, during interaction with exogenous brain antigens. Abscissa 1, stress-exposed animals; 2, animals preinjected with BHT (120 mg/kg); 3, animals immunized with brain extract; 4, animals immunized with brain extract after injection of BNT (120 mg/kg)

As can be seen from Table 3, the factor of stress resulted in the activation of LPO and a simultaneous inhibition of the specific binding of [^{3}H] dihydroalprenolol by B-adrenoreceptors. Similar results were obtained in experiments with brain cortex membranes in vitro (Fig. 2). In these model experiments with the Fe^{2+} + ascorbate system, it was demonstrated that stress-induced changes in the properties of the receptor proteins could be simulated by LPO activation in brain cortex membranes in vitro.

The next task was to evaluate the pathophysiological significance of the stress-induced LPO in rat brain tissues, with a special emphasis on the integral parameters of the physiological state of nervous tissues and, in particular, on the functional activity of the retina, a peripheral part of the visual analyzer of the brain. The functional activity of the retina was registered as an electroretinogram (Shvedova et al. 1982). The data from Table 4 suggest that 2 h after the stress exposure, the amplitudes of the a and b waves were sharply decreased and returned to the normal level only after 72 h. In this case as well, a marked protective effect was exerted by BHT.

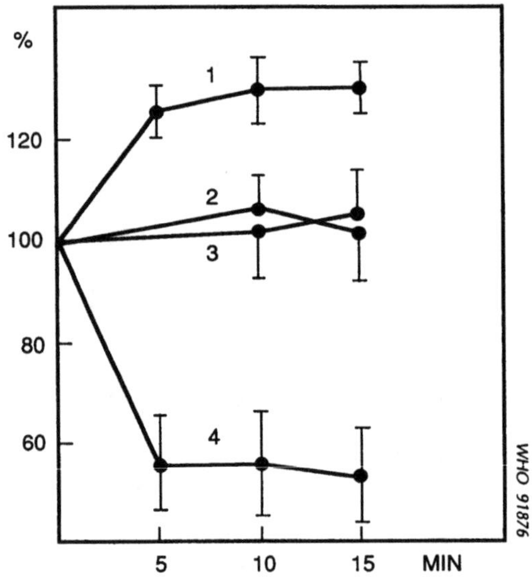

Fig. 2. In vitro effects of LPO on the binding of ^{3}H-serotonin and ^{3}H-diazepam in the membrane fraction of a rat brain cortex. Ordinate, binding of labelled ligans (% of control). Abscissa incubation time in minutes. 1,4, without BHT; 2,3, with BHT (5.10^{-5}M); 1,3, ^{3}H-diazepam; 2,4, ^{3}H-serotonin

Table 4. Effect of stress on electroretinograms of rats and protective action of BHT (mV)

Time elapsed after stress exposure	Experimental group			
	Control		BHT (120 mg/kg)	
	a	b	a	b
Control	267±17	625±47	268±17	626±47
2 h	109±8[a]	256±18[a]	175±17[a]	510±28[b]
24 h	120±10[a]	300±23[a]	260±19	617±34
48 h	130±20[a]	340±37[a]	260±18	620±35
72 h	260±30	630±50	–	–

[a] $p<0.001$; [b] $p<0.05$

We may therefore conclude that the accumulation of endogenous LPO products under stress is accompanied by an aggravation of the functional activity of the brain, an increase in the permeability of the hematoencephalic barrier membranes and changes in the properties of membrane-bound proteins. In this connection it seemed essential to extrapolate these results to the human organism. For this purpose we attempted to evaluate the potentials of the LPO activation in humans exposed to emotional stress.

The stress in the form of an operator activity was evoked in volunteers under conditions of a predefined time deficiency. The intensity of LPO was studied by a noninvasive method modified for a human organism. This method consists of the measurement of the contents of pentane, a volatile end product of LPO, in the

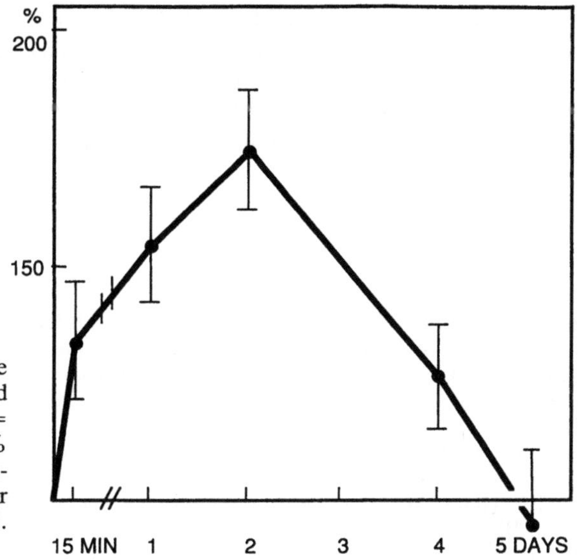

Fig. 3. Dynamics of the pentane content in the expired air induced by emotional stress in man (n = 13). Ordinate, pentane content (% of control before the stress exposure). Abscissa, time elapsed after the stress exposure (Prilipko et al. 1982)

expired air (Prilipko et al. 1982). It was found that, in a human organism, LPO activation occurred within the first 2 days after the stress exposure (Fig. 3).

When comparing these results with the above-mentioned data, it would be natural to assume that LPO is a constitutive reaction of the general pathophysiological process occurring in some patients with mental disorders, in particular, those with schizophrenia (Prilipko and Lideman a, b 1982). Using a polarographic technique, it was shown that the level of LPO products in the blood of such patients was considerably increased (Table 5). The levels of primary and secondary LPO products in this case were correlated with the degree of the disease progression, thus indicating LPO involvement in the pathophysiological process.

Table 5. The content of lipid hydroperoxides and end products of LPO in the peripheral blood of schizophrenic patients and healthy donors

Experimental groups	Number of individuals tested	Lipid hydroperoxides (nmol/mg lipid)	LPO end products (nA/mg lipid)
Healthy donors	25	0.72±0.22	18.0±2.4
Males	12	0.57±0.24	20.7±5.0
Females	13	0.85±0.37	15.6±0.9
Schizophrenic patients	23	5.02±1.10	54.8±10.3
Males	9	5.21±1.47	50.4±12.9
Females	14	4.83±1.60	55.8±10.3
Nuclear schizophrenia	9	8.12±2.40	79.9±22.6
Shift-like schizophrenia	11	2.74±0.30	35.0±5.1

In conclusion, it may be said that LPO plays a crucial role in the stress-induced damage of brain membranes and permits the assessment of the involvement of these processes in the pathogenesis of schizophrenia. At this point, free radical scavengers seem to be a useful tool in the prophylaxis of stress related damages and may be used in the future in the development of a comprehensive therapy of endogenous psychoses.

References

Bennett, IP, Salvatore IE, Bylund DB, Gillin IC,Wyatt RI, Snyder SH (1979) Neurotransmitter receptors in frontal cortex schizophrenics. Arch Gen Psychiatry 36 7: 927–934

Bogdanova ED, Kagan VE, Kuliev IY, Meerson FZ, Prilipko LL (1981) Activation of lipid peroxidation in the brain and production of antibody against brain antigens in stress. Immunologiya 2: 65–66

Desiderato, O, Testa M (1976) Shock-stress, gastric lesions and habituation in chronic gastric fistula rat. Physiol Behav. 16: 67–73

Henn FA, Henn SW (1981) Phospholipids as markers for schizophrenia. In: Usdin E, Hanin I (eds) Biological markers in psychiatry and neurology. Pergamon, New York, pp 183–186

Hitzemann RI, Garver DL (1981) Abnormalities in membrane lipids associated with deficiencies in lithium counterflow. In: Usdin E, Hanin I (eds) Biological markers in psychiatry and neurology. Pergamon, New York, pp 177–182

Kagan VE, Prilipko LL, Savov VM, Pisarev VA, Eluashvili IE, Kozlov YuP (1979) Participation of oxygen activated species in enzymatic lipid peroxidation in biomembranes. Biokhimiia 44 3: 482–489

Lee T, Seeman P (1981) Brain dopamine receptors in schizophrenia. In: Usdin E, Hanin I (eds) Biological markers in psychiatry and neurology. Pergamon, New York, pp 219–226

Leller EA, Davis JM (1980) Protein content and monoamine oxidase activity in platelets. Schizophr Bull. 11 3: 267–274

Libe ML, Bogdanova ED, Rozenberg AE, Prilipko LL, Kagan VE, Kozlov Yu P (1981) Binding of 3H-serotonin, 3H-diazepam and lipid peroxidation in brain membranes. Bull Exp Biol Med (USSR) II: 552–554

Lideman RR (1972) On the question of membranotropic factor of schizophrenic patients blood serum. J Nevropatol Psychiatrii S.S. Korsakova (USSR) 72 4: 604–612

Lohr JB, Cadet JL, Lohr M, Larson L, Wasli E, Wade L, Hylton R, Vidoni C, Jeste DV, Wyatt RJ (1988) Vitamin E in the treatment of tardive dyskinesia: the possible involvement of free radical mechanisms Schizophr Bull 14 2: 291–296

Prilipko LL, Lideman RR (1982a) Lipid peroxidation as a causative factor in membrane-bound protein modification in nervous cells of schizophrenic patients. Vestn Akad. Med. Nauk SSSR 1: 33–36

Prilipko LL, Lideman RR (1982b) Peripheral blood lymphocytes as biological indicators and pathophysiological processes in schizophrenia. J. Nevropatol Psichiatrii S. S. Korsakova (USSR) 1: 79–86

Prilipko LL, Orlov ON, Ivanova SM, Kagan VE, Meerson FZ, Ushakov AS (1982) Activation of lipid peroxidation in humans under stress as measured by the pentane content in the expired air. Dokl Akad Nauk SSSR 265 4: 1010–1013

Shevedova AA, Kagan VE, Kuliev I Ya, Dobrina SK, Prilipko LL, Meerson FZ, Kozlov Yu P (1982) Lipid peroxidation and damage of the retina in rats exposed to stress. Bull Exp Biol Med (USSR) 4: 24–26

Wise CD, Baden MM, Stein L (1974) Post-mortem measurements of enzymes in human brain: evidence of a central noradrenergic deficiency in schizophrenia. J. Psychiatr Res 11 3: 185–198

Zeeberg BR, Gibson RE, Reba RC (1990) Elevated D^2 dopamine receptors in drug-naive schizophrenics. 239: 789

Side-Effects of Psychotropic Therapy

R. J. Sram and *B. Binkova*

It has long been hypothesized that schizophrenia may be related to a specific injury of genetic material. There is some evidence of chromosomal abnormalities (DeLisi et al. 1988). Another study examined the repair defects seen in schizophrenic patients when their lymphocytes were treated with 4-nitroquinoline or gamma-radiation (Zasukhina 1987). However, no significant DNA-repair defect was observed by Magin et al. (1991).

Several papers indicating the side-effects of long-term therapy with neuroleptics, especially those having the structure of phenothiazines, have been published recently. Patients treated with neuroleptics had a higher level of lipid peroxidation (LPO) in the cerebrospinal fluid (Pall et al. 1987). This finding indicates possible injury to neuronal membranes. Similarly, the results of treatment of some tardive dyskinesia (TD) symptoms with high doses of alpha-tocopherol support the idea that the induction of free radicals by long-term neuroleptic therapy can be suppressed by antioxidants (Cadet and Lohr 1987).

Oxygen radicals, especially reactive hydroxy radicals, that cannot be eliminated from cells by a protective enzymatic system may act as mutagens, promotors and carcinogens. They initiate the chain reaction of LPO acting the structural degradation of biomembranes, as well as oxidative damage of proteins, RNA and DNA. The activation of these oxidative processes may induce gradual cellular degeneration and damage (Pero et al. 1989).

Current knowledge indicates the significance of oxidative damage to the human organism and the mutual connection of the LPO and DNA repair mechanisms as processes which may speed up aging, atherosclerosis, cancer, etc. (Ames 1983). This oxidative damage may be suppressed by antioxidants acting as free radical scavengers (Sram et al. 1986; De Flora and Ramel 1988; Ames 1989; Dowson 1989). This is the reason why we studied oxidative damage in schizophrenic patients and the effect of neuroleptics in vitro and in vivo.

Schizophrenic Patients

Subjects for this study were chosen from inpatients who were treated continually with phenothiazines for at least one year and were also hospitalized for at least one year (the dg in schizophrenics was determined approximately 28 years earlier). An increased frequency or AB.C. and LPO and a decreased level of

UDS and vitamin C (Table 1) were observed. These results correspond to the hypothesis about the effect of neuroleptics on free radicals and catecholamine metabolism (Topinka et al. 1988).

It was therefore suggested that cell membrane destabilisation by chronic use of neuroleptics, resulting in the persistent form of TD, might be effectively compensated by alpha-tocopherol (Cadet and Lohr 1987). Our pilot study analysed the effect of 6 weeks of therapy with a daily dose of 600 mg of alpha-tocopherol (Table 2). Administration of vitamins decreased LPO (Yagi 1976; Bradford 1976) in plasma ($p<0.01$) as well as in lymphocytes ($p<0.05$) and

Table 1. Effects of neuroleptics on chromosome aberrations, UDS and LPO levels in schizophrenics

Group	Schizophrenics	Controls
Number	24	23
AGE	48.1	38.3
s.d.	11,8	6.0
Analysed cells	2 400	2 300
% AB.C.	2.60[a]	1.52
B/C	0.037	0.016
UDS (T/C)	2,95[a]	3,63
s.d.	0,91	0,85
LPO Plasma (nmol MDA/ml)	1,41[b]	1,09
s.d.	0,22	0,13
Ascorbic acid (mg/l)	5,3	10,1
s.d.	3,5	3,6

[a] $p < 0,05$, [b] $p < 0,01$

Table 2. Effect of α-tocopherol on UDS, LPO and vitamin levels in patients with tardive dyskinesia (N=13)

	Supplementation with α-tocopherol	
	Before	After
UDS (T/C)	3,16	4,24[a]
s.d.	0,68	0,91
LPO Plasma (nmol MDA/ml)	1,77	1,32[a]
s.d.	0,18	0,17
LPO Lymphocytes (nmol/mg)	1,65	1,29[b]
s.d.	0,34	0,22
Ascorbic acid (mg/l)	6,3	6,1
s.d.	4,6	4,8
α-tocopherol (mg/l)	9,8	17,4[a]
s.d.	3,3	9,6

[a] $p < 0.01$; [b] $p < 0,01$

increased UDS (Martin et al. 1978) ($p<0.01$) and the plasma level of alpha-tocopherol (Driskell 1982) ($p<0.01$).

Based on the Brief Psychiatric Rating Scale (BPRS), schizophrenics improved after administration of vitamins in all factors (anxiety-depression $p<0.05$; activation $p<0.01$; hostile suspiciousness $p<0.01$) and slightly improved on the Clinical Global Impression Scale (CGI, $p<0.05$). Comparing the UDS and LPO levels for schizophrenics with and without TD, we found that the injury was more pronounced in patients with TD. It seems that a longer period of antioxidant therapy is necessary to analyse more thoroughly its effect on free radicals, especially on the clinical symptoms. We need to clarify whether the suppression of the side-effects of neuroleptics is not simultaneously reducing the effect of the neuroleptics therapy.

Neuroleptics in vitro

Lymphocytes were isolated from heparinized blood of healthy volunteers (Harris and Ukaejiofo 1970). The effect of chemicals on free radical processes was investigated by treatment with the oxidant and/or antioxidant. This step was followed by adding Fe^{2+}/ascorbate or H_2O_2 to activate molecular oxygen. DNA oxidative damage was measured as unscheduled DNA synthesis and oxidative damage of biomembranes was measured as the level of lipid peroxidation by the TBA assay (Ohkawa et al. 1979) (Fig. 1).

We have studied the effect of several neuroleptic drugs (Fig. 2) in our in vitro system, in the concentration range of 1 μmol/l to 0.1 mmol/l. As can be seen from the results on rat brain synaptosomes (Hajos 1975), chlorpromazine and perchlorperazine exhibited an antioxidant effect at higher concentrations, whereas the other two drugs had no effect on the in vitro induced synaptosomal LPO level (Fig. 3). In human lymphocytes all these drugs significantly increased the LPO level (Fig. 4). The two phenothiazine compounds – chlorpromazine and perchlorperazine – exhibited a decrease in these changes at increasing concentrations. Chlorpromazine, meclopine and haloperidol significantly potentiated the free radical damage of DNA initiated by Fe^{2+}/H_2O_2 (Fig. 5). When we compare the results on lymphocytes and synaptosomes, the differences are probably due to the participation of these drugs in free radical production during metabolism by lymphocytes.

The well-known effective antioxidant and membrane stabilizer, alpha-tocopherol, very efficiently prevented free radical damage induced in the presence of neuroleptics in these cells when the suspension of cells was pretreated with alpha-tocopherol in a final concentration of 200 μmol/l (Figs. 6 and 7).

Neuroleptics in vivo

To verify the in vitro results, we also analysed the effects of a model neuroleptic in vivo (Sram et al. 1990). Male Wistar strain rats, aged 6 months, were

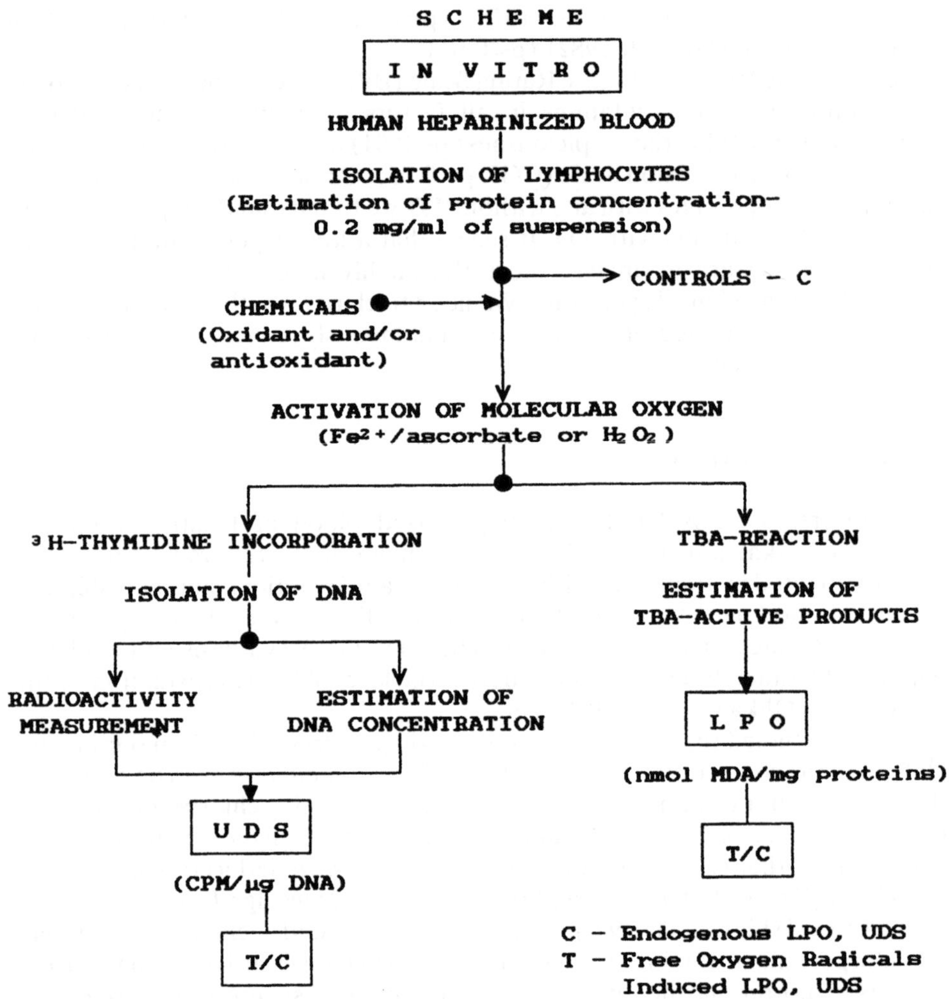

Fig. 1. Scheme of UDS and LPO analysis in vitro (oxidative damage induced by free radicals generating system)

administered one of the neuroleptic drugs and vitamin E, separately and together, during a 12-month period. We chose the depot neuroleptic fluphenazine, which was injected intramuscularly into the rats every 3 weeks. The depot form of drug gave rise to a long-acting and stable serum level. The effects of treatment were analysed after 3, 6 and 12 months (Fig. 8).

The heparinized blood was collected from the heart. After isolation of peripheral lymphocytes, UDS induced by the standard concentration of MNNG was estimated. The brain was dissected into cerebral cortex, hippocampus and hypothalamus, where the endogenous LPO levels were measured using a TBA assay (Fig. 9). Statistically significant changes in comparison to the control

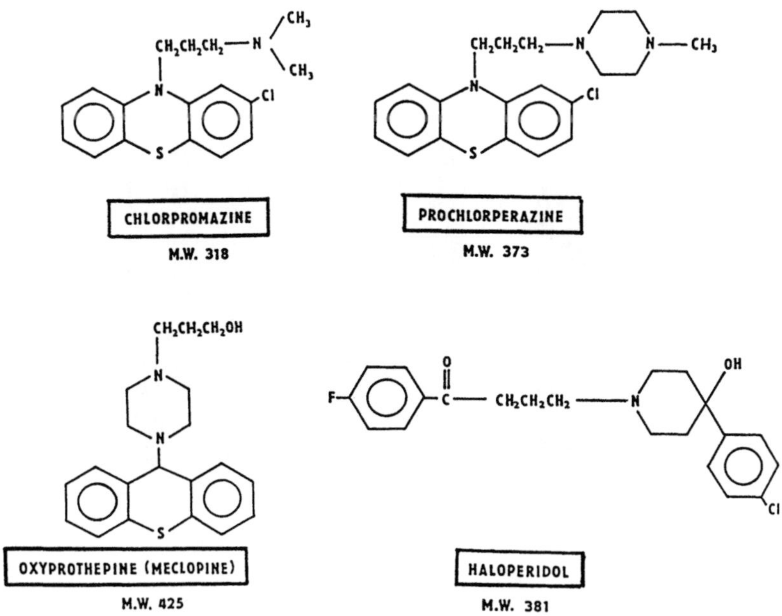

Fig. 2. Chemical structure of studied neuroleptics

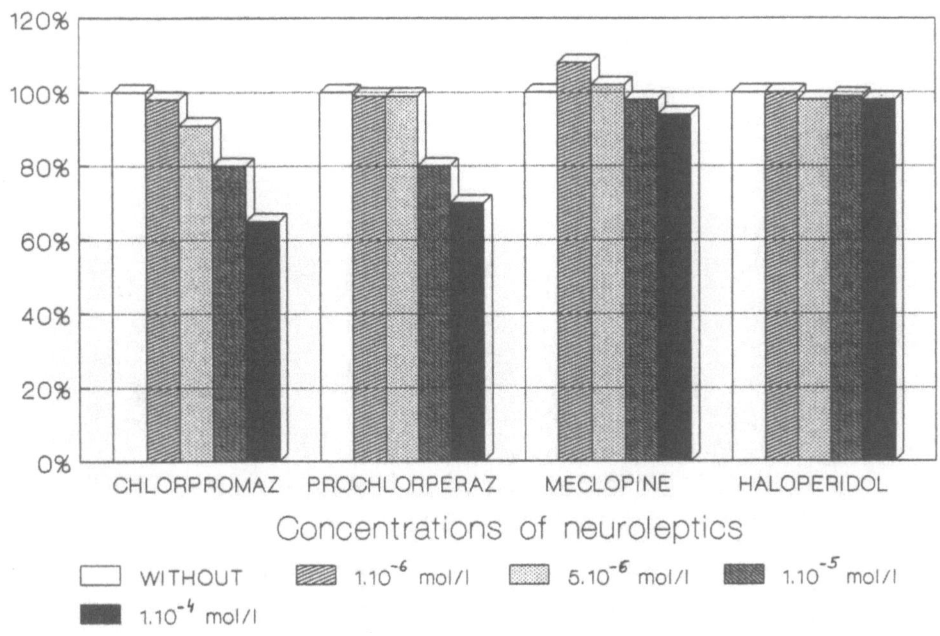

Fig. 3. Effect of neuroleptics on LPO. In vitro test – rat synaptosomes

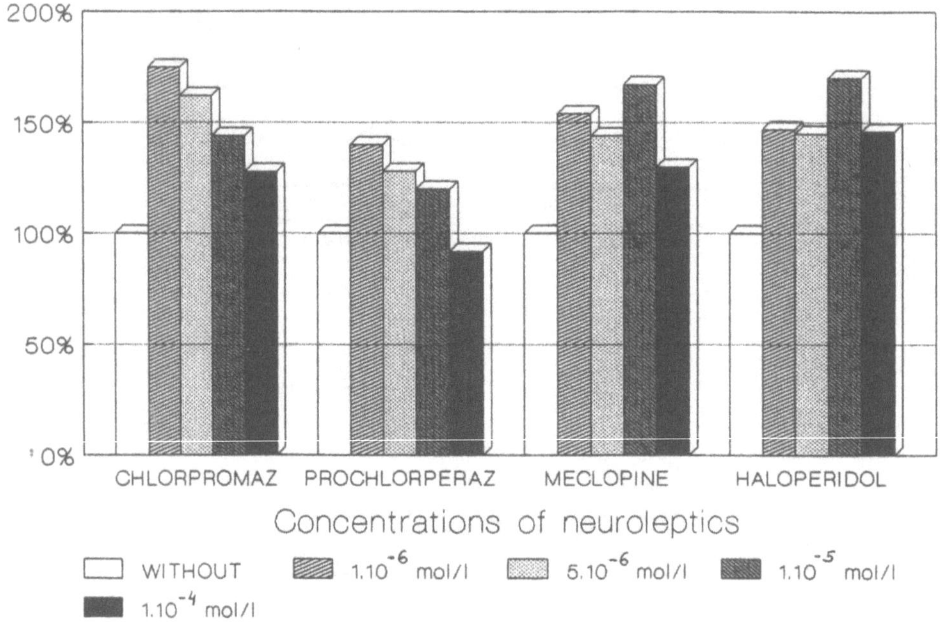

Fig. 4. Effect of neuroleptics on LPO. In vitro test – human lymphocytes

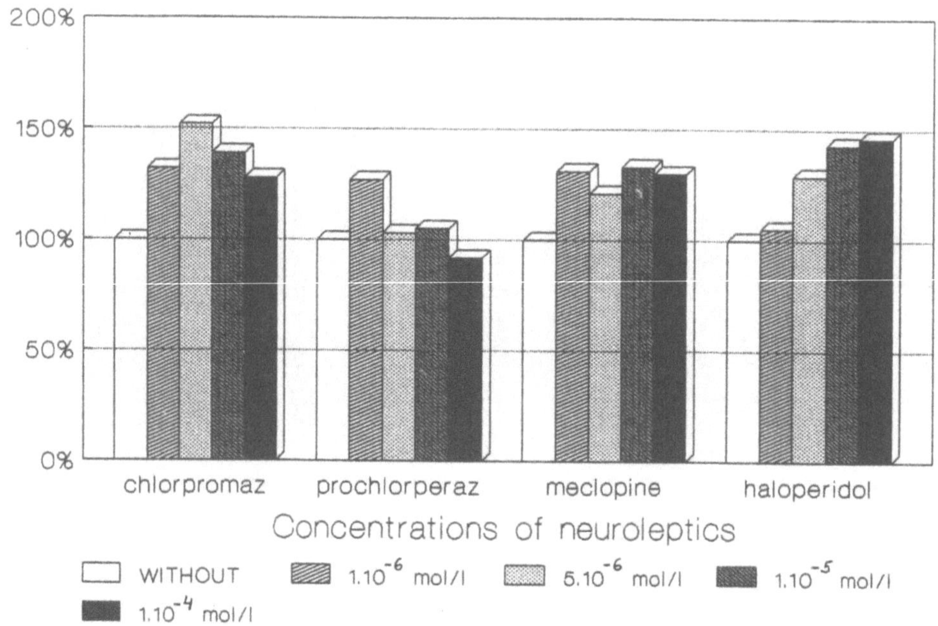

Fig. 5. Effect of neuroleptics on UDS. In vitro test – human lymphocytes

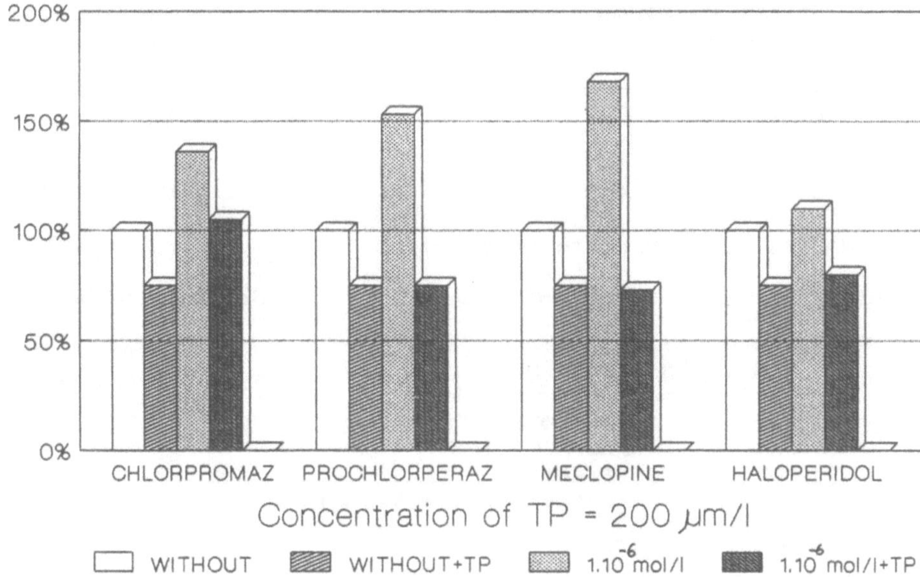

Fig. 6. Effect of alpha-tocopherol on LPO induced in vitro and by neuroleptics. In vitro test – human lymphocytes

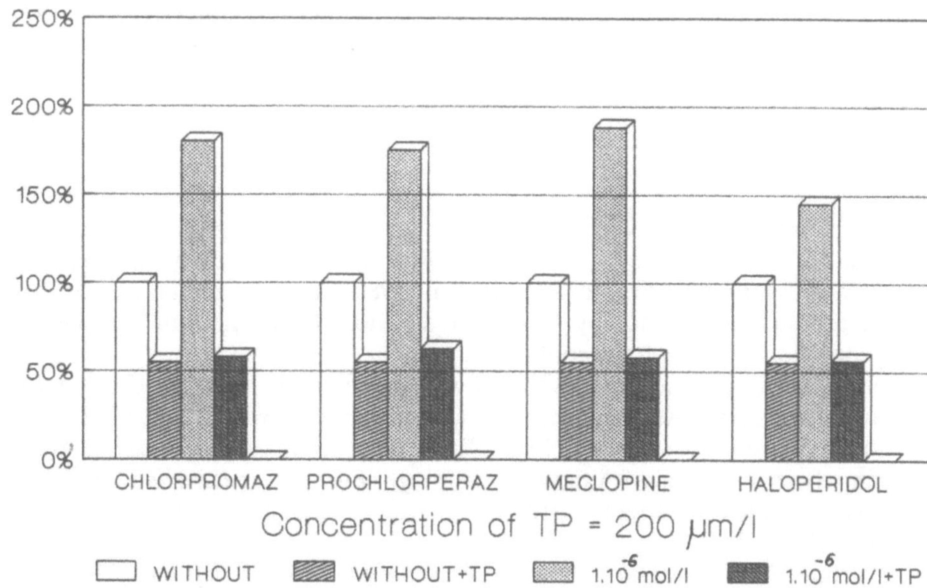

Fig. 7. Effect of alpha-tocopherol on UDS induced in vitro and by neuroleptics. In vitro test – human lymphocytes

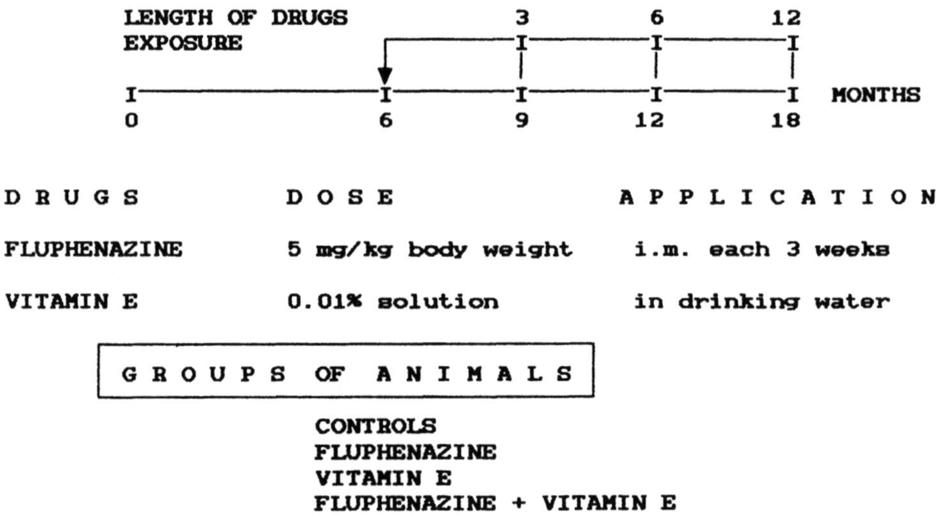

Fig. 8. Scheme of experiments on rats

group of the same age (10 animals were analysed in each group) were observed only after 6 and 12 months of treatment. Exposure to fluphenazine decreased DNA excision repair and increased LPO in cerebral cortex and hypothalamus. Simultaneous administration of alpha-tocopherol with the neuroleptic drug eliminated these changes. Therefore, it may be assumed that simultaneous administration of alpha-tocopherol can suppress some side-effects of long-term drug treatment and may stimulate the therapeutic approach of neuroleptics (Table 3).

Effect of antioxidants

In our in vitro system with the human peripheral lymphocytes we also investigated some chemical and natural substances in which an antioxidant activity was assumed to exist (Topinka et al. 1989). Gingko biloba extract 761 (EGb), produced in France by IPSEN (De Feudis 1991), had an antioxidant effect comparable to alpha-tocopherol.

No significant influence on LPO-level was observed by nootropic drugs, such as pyritinol and cetrexine, in which the nootropic activity was often explained by the scavenging of free radicals (Fig. 10). Similar results were obtained from analysing their effect on DNA oxidative damage (Fig. 11). In this case riboflavin also has a significant weakly protective effect against free radicals DNA damage.

The properties of Gingko biloba extract were studied in much more detail to compare its activity with the most efficient membrane antioxidant, alpha-tocopherol, in the following areas: the dependence of LPO inhibition in rat brain synaptosomes (Fig. 12), in human lymphocytes (Fig. 13) and the effect on DNA

Table 3. Effect of alpha-tocopherol on USD in peripheral lymphocytes and LPO in cortex and hypothalamus after 3, 6 and 12 months administration of fluphenazine to 6 months old rats (groups N = 10)

	Groups		Time of Administration 3 months	6 months	12 months
UDS (T/C) peripheral lymphocates	controls	$\bar{x}$	1.90	1.87	1.97
		s.o.	0.15	0.24	0.35
	fluphenazine	$\bar{x}$	1.60[b]	1.43[b]	1.62[b]
		s.o.	0.14	0.29	0.21
	vitamin E	$\bar{x}$	1.93	1.80	2.59
		s.o.	0.10	0.22	0.42
	fluphenazine +	$\bar{x}$	1.68	1.65	2.09
	+ vitamin E	s.o.	0.29	0.31	0.27
LPO (nmol MDA/g tissue) in cerebral cortex	controls	$\bar{x}$	27.9	28.8	30.4
		s.o.	4.4	4.2	1.5
	fluphenazine	$\bar{x}$	31.4	36.0[a]	33.0[b]
		s.o.	2.7	6.2	2.3
	vitamin E	$\bar{x}$	30.3	31.8	30.9
		s.o.	4.9	2.6	2.3
	fluphenazine +	$\bar{x}$	27.4	31.7	29.6
	+ vitamin E	s.o.	3.7	3.2	2.7
LPO (nmol MDA/g tissue) in hypothalamus	controls	$\bar{x}$	28.6	28.2	31.3
		s.o.	3.5	4.5	1.9
	fluphenazine	$\bar{x}$	30.0	33.8[a]	34.0[a]
		s.o.	3.6	5.1	2.4
	vitamin E	$\bar{x}$	28.3	29.1	30.4
		s.o.	4.2	2.4	1.7
	fluphenazine +	$\bar{x}$	27.6	29.6	31.6
	+ vitamin E	s.o.	1.8	6.2	4.3

[a] $p < 0.05$ (to controls)
[b] $p < 0.01$

oxidative damage (Fig. 14). It is evident from these results that EGb and alpha-tocopherol have a similar inhibitory effect on lipid peroxidation in both rat synaptosomes and human lymphocytes. The higher protective effect of alpha-tocopherol against DNA-oxidative damage may be connected with its stabilizing effect on membrane lipid bilayer. On the other hand, only 47 % of the total EGb represented the antiradical active compounds such as flavonoids (Pincemail and Deby 1988; Wall et al. 1990).

We also studied the effects of three months of supplementation of EGb and alpha-tocopherol to old rats at the age of 24 months (Fig. 15), because it has been proposed that oxygen radicals play an important role in the aging process. Both substances were administered to the rats in drinking water at a dose of 100 mg/kg body weight. As may be seen from the results, both substances had a significantly positive influence on endogenous LPO levels in cerebral cortex and hippocampus in comparison with the control, non-supplemented group of aged rats.

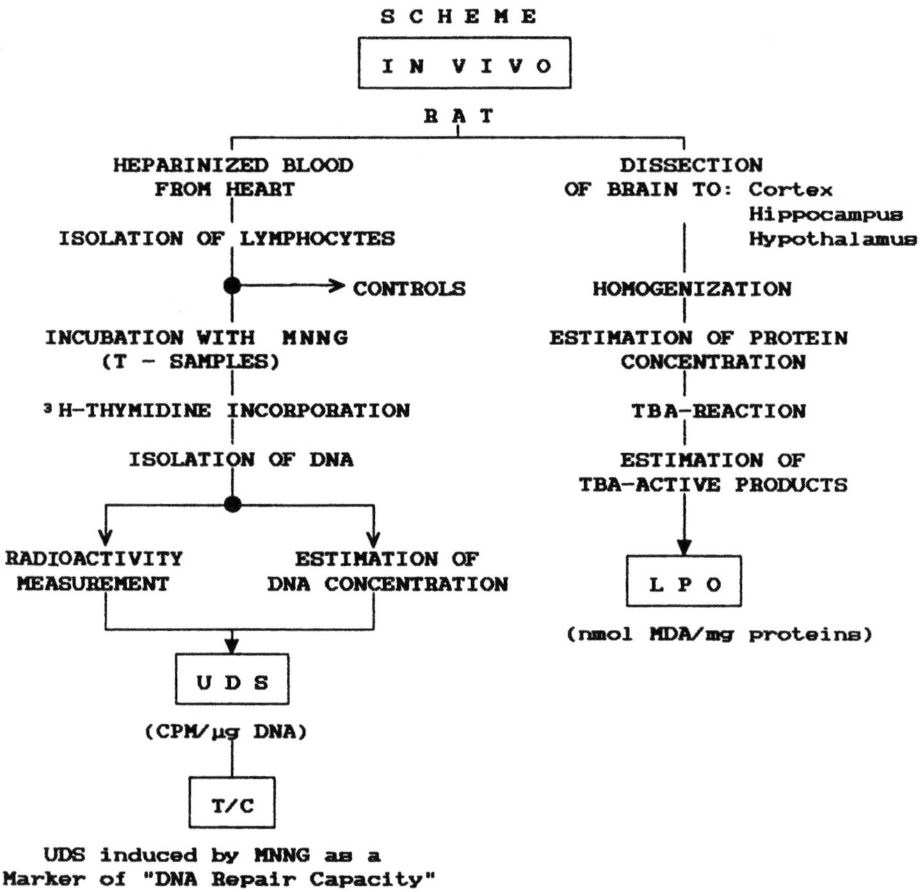

Fig. 9. Scheme of UDS and LPO analysis in vivo

Conclusion

Summing up our results, we believe that the long-term use of antioxidants may become a promising therapy in free radical diseases. Experimental data, as well as experience with their long-term application in humans, support the idea that pharmacological interventions by antioxidants may be efficient in the prevention of dieseases associated with oxidative stress (such as aging, atherosclerosis) or in the elimination of some drug side-effects, as may be seen for neuroleptics. These studies suggest that simultaneous treatment with antioxidants and certain neuroleptic drugs will produce a beneficial effect by reducing free radical levels.

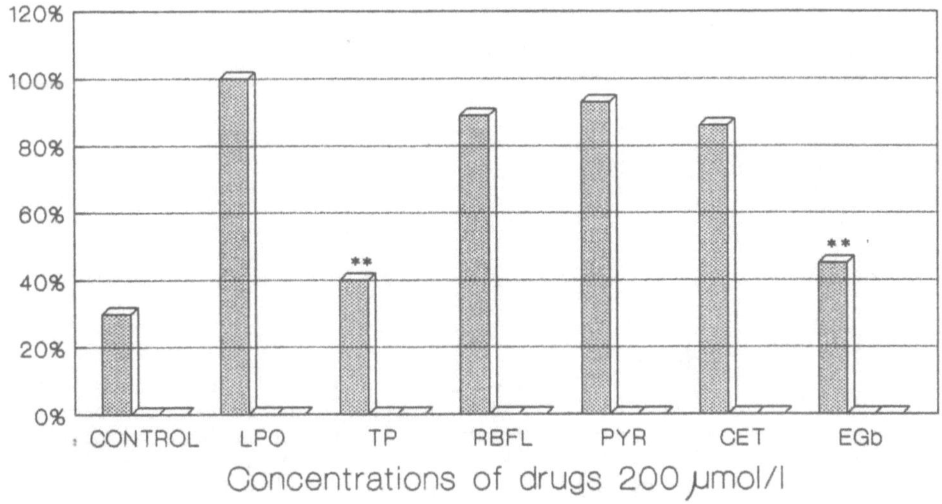

TP=Alpha-tocopherol; RBFL=Riboflavin;
PYR=Pyritinol; CET=Cetrexin;
EGb= Ginkgo biloba 761

Fig. 10. Effect of antioxidants on LPO. In vitro test

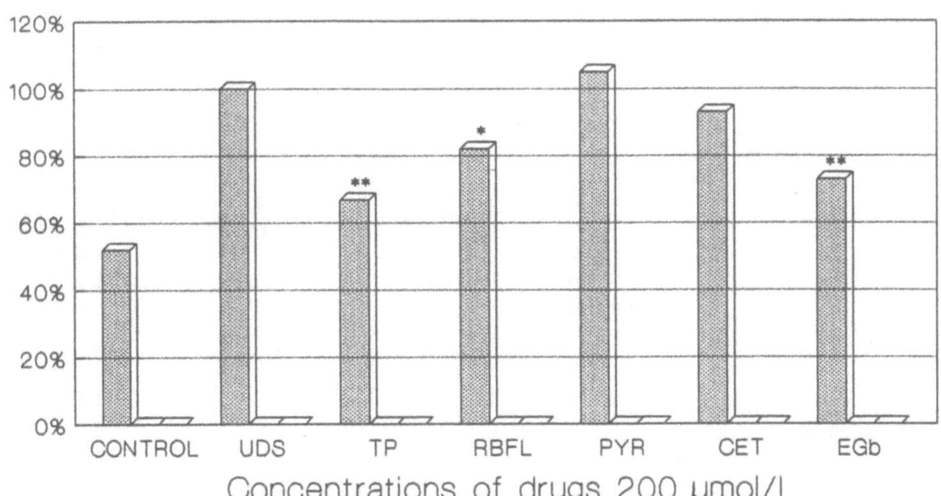

TP=Alpha-tocopherol; RBFL=Riboflavin;
PYR=Pyritinol; CET=Cetrexin;
EGb= Ginkgo biloba 761

Fig. 11. Effect of antioxidants on UDS. In vitro test

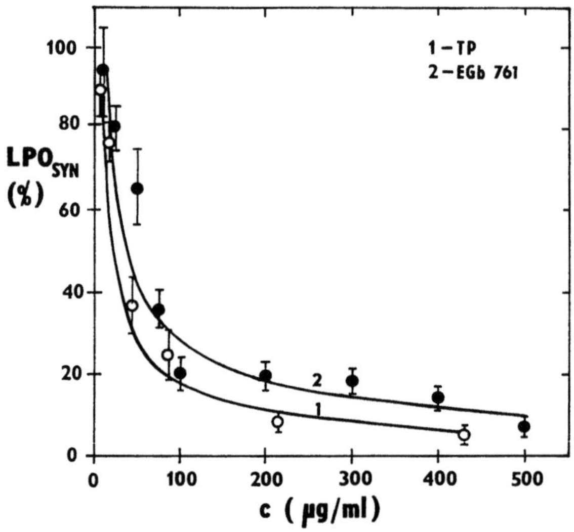

Fig. 12. Inhibitory effect of EGb 761 and TP on the lipid peroxidation in rat brain synaptosomes (LPO_{LYM}) induced by Fe^{2+} (3 μmol/l) and H_2O_2 (1 mmol/l)

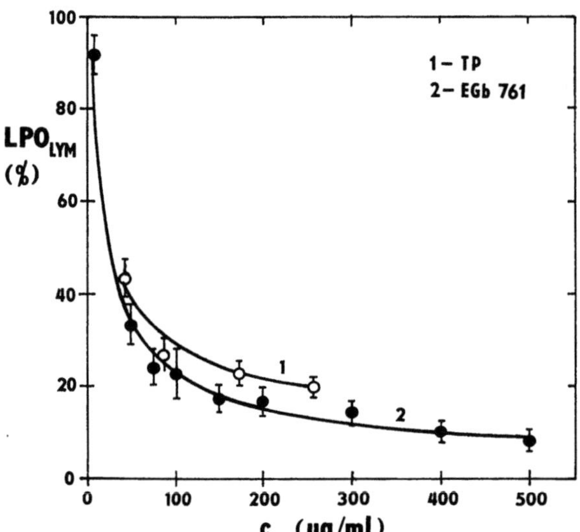

Fig. 13. Inhibitory effect of EGb 761 and TP on the lipid peroxidation in human lymphocytes (LPO_{LYM}) induced by Fe^{2+} (3 μmol/l) and H_2O_2 (1 mmol/l)

References

Ames BN (1983) Dietary carcinogens and anticarcinogens (oxygen radicals and degenerative-diseases). Science 221: 1256–1264

Ames BN (1989) Endogenous DNA damage as related to cancer and aging. Mutation Res 214: 41–46

Bradford MM (1976) A rapid method for the quantitation of microgram quantities of protein utilizing the principle of protein-dye binding. Anal Biochem 72: 248–254

Cadet JL, Lohr JB (1987) Free radicals and the development pathobiology of schizophrenic burnout. Integr Psychiatry 5: 40–48

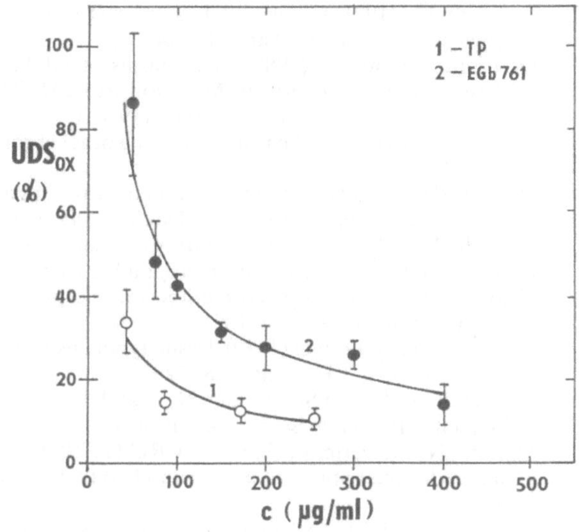

Fig. 14. Inhibitory effect of EGb 761 and TP on the lipid peroxidation in human lymphocytes (LPO_{LYM}) induced by Fe^{2+} (10 µmol/l) and H_2O_2 (1 mmol/l)

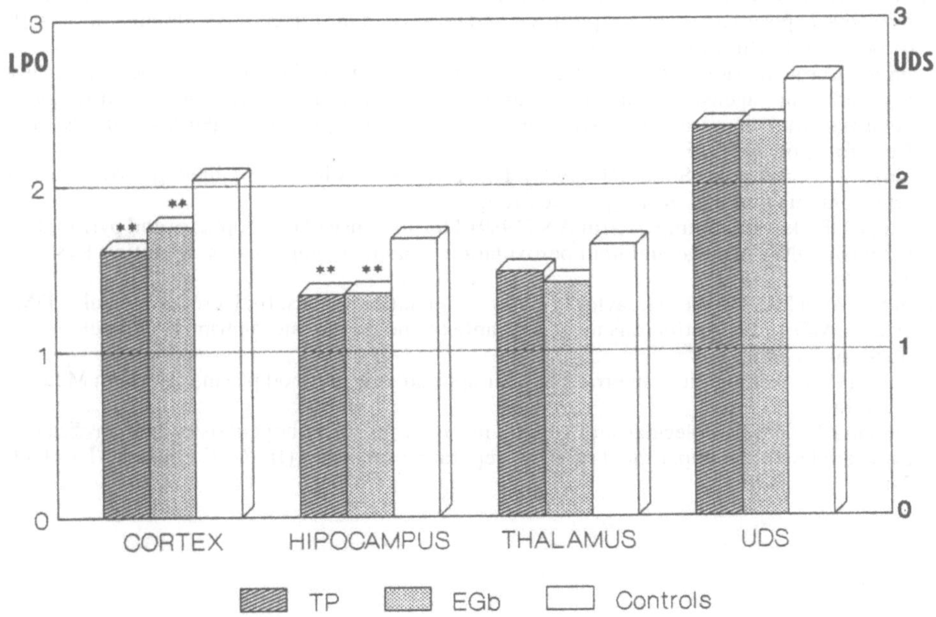

Fig. 15. Effect of TP and EGb-761 on LPO and UDS. 3 months supplementation to old rats

De Feudis FV (1991) Gingko biloba extract (EGB 761): Pharmacological activities and clinical applications. Elsevier, Paris, 187 pp
De Flora S, Ramel C (1988) Mechanisms of inhibitors of mutagenesis and carcinogenesis. Classification and overview. Mutation Res 202: 285–306
De Lisi L, Reiss A, White B, Gershon E (1988) Cytogenetic studies of males with schizophrenia: screening for fragile-X chromosome and other chromosomal abnormalities. Schizophr Res 1: 277–281
Dowson JH (1989) Drug treatment for cognitive impairment due to aging and disease: current and future strategies. Int J Ger Psych 4: 345–353
Driskell JW (1982) Measurement of vitamin A and vitamin E in human serum by high-performance liquid chromatography. J Chromatogr 231: 439–444
Hajos F (1975) An improved method for the preparation of synaptosomal fraction in high purity. Brain Res 93: 185–189
Harris R, Ukaejiofo RO (1970) Tissue typing using a routine one-step lymphocyte separation technique. Brit J Haematol 18: 229–235
Magin GK, Robinson SH, Breslin N, Wjall RJ, Alexander RC (1991) DNA repair and mutant frequency in schizophrenia. Mutation Res 266: 241–246
Martin CN, McDermid AC, Garner RC (1978) Testing of known carcinogens and noncarcinogens for their ability to induce unscheduled DNA synthesis in HeLa cells. Cancer Res 38: 2621–2627
Ohkawa H, Ohishi N, Yagi K (1979) Assay for lipid peroxides in animals tissues by thiobarbituric acid reaction. Anal Biochem 95 351–358
Pall HS, Blace DR, Williams AC, Lunec J (1987) Evidence of enhanced lipid peroxidation in the cerebrospinal fluid of patients taking phenothiazines. Lancet ii: 596–599
Pero RW, Anderson M, Markowitz M, Doyle G, Hunnicutt C (1989) Suppression of DNA repair by oxidative stress. Proc Am Ass Canc Res 30: abstract No. 7938
Pincemail J, Deby C (1988) The antiradical properties of Gingko biloba extract. In: Funfgeld EW (ed) Rokan-Gingko biloba. Recent results in pharmacology and clinic. Springer Verlag, Berlin, pp 71–82
Sram RJ, Binkova B, Topinka J, Fojtikova I (1990) Inhibition of DNA repair synthesis in the rat by in vivo exposure to psychotropic drugs and reversal of the effect by co-administration with α-tocopherol. Mutation Res 244: 331–335
Sram RJ, Cerna M, Hola N (1986) Effect of ascorbic acid prophylaxis in groups occupationally exposed to mutagens. In: Ramel C, Lambert B, Magnusson J (eds) Genetic toxicology of environmental chemicals. Part B: Genetic effects and applied mutagenesis. Alan R Liss, New York, pp 327–335
Topinka J, Binkova B, Sram RJ (1988) DNA repair synthesis and lipid peroxidation in schizophrenic patients. Schizophr Res 1: 152
Topinka J, Binkova B, Sram RJ, Erin AN (1989) The influence of α-tocopherol and pyritinol on oxidative DNA damage and lipid peroxidation in human lymphocytes. Mutation Res 225: 131–136
Wall Me, Wani MC, Hughes TJ, Taylor H (1990) Plant antimutagens. In: Kuroda Y, Shankel DM, Waters MD (eds) Antimutagenesis and anticarcinogenesis mechanism II. Plenum, New York, pp 61–78
Yagi K (1976) A simple fluorometric assay for lipoperoxide in blood plasma. Biochem Med 15: 212–216
Zasukhina G (1987) Molecular and genetic mechanisms of pathogenesis of human diseases associated with the impairment of DNA reparative processes (In Russian). Arkh Patol 49: 3–9

Lipid Peroxidation and Antioxidants in Childhood Epilepsy

A.O. Ogunmekan and *P.A. Hwang*

Summary

Evidence linking vitamin E deficiency with neurological abnormalities indicates that vitamin E may have an important central nervous system role relating to membrane stability and function. In an experimental model, vitamin E (α-tocopheryl acetate) reduced seizure activity. Vitamin E inhibits the effects of oxidation in brain tissues (antioxidant) and is a free radical scavenger. Free radical scavengers have been implicated in the epileptic brain injury caused by prolonged seizures due to hyperoxia. Vitamin E is a membrane stabilizer protecting membranes from peroxidation damage. Clinical studies in children have shown reduced plasma levels of vitamin E in seizure patients receiving antiepileptic drugs (AEDs). An uncontrolled, non-blind clinical trial suggested that add-on vitamin E improved seizure control in some patients. Since vitamin E deficiency hypothetically exacerbates epileptic seizures in patients receiving AEDs, treatment of the vitamin E deficiency should improve seizure control and/or neurological function. A prospective, modified crossover, randomized, double-blind, placebo-controlled clinical trial of add-on vitamn E therapy (400 IU/day) was then carried out in 24 epileptic patients who were refractory to AEDs. There was a significant reduction of seizures in 10 of 12 patients. This result was significantly different from that with controls given placebo (0 of 12, $p < 0.05$). This result did not appear to be due to the effects of change in the plasma levels of the co-medication. There were no adverse side effects. This therapeutic observation accords with previous reports and some experimental studies. This finding may be very important to the 10%–20% of the epileptic population whose seizures are inadequately controlled with the best anticonvulsant drugs presently available.

Introduction

Vitamin E is well accepted as the major fat-soluble, chain-breaking antioxidant in the body. It is a well-known free radical scavenger, protecting cell membranes from peroxidation damage. Evidence linking vitamin E deficiency with neurological abnormalities indicates that vitamin E may have an important central nervous system role relating to membrane stability and function. Certain age-related central nervous system changes, such as senile dementia and Parkinson's disease, may result from free radical-related damage to cell membranes. Research indicates that vitamin E may have an important role in the neurobiology of aging and that the requirement for vitamin E by the central nervous system may be altered during aging.

Vitamin E remains an enigma and a continuing challenge to both nutritional and clinical scientists. The subcellular role of vitamin E has been particularly difficult to establish, although it clearly functions as a biologic antioxidant. Provocative experimental results in free radical biology have stimulated ever-increasing interest during recent years in the antioxidant capability of vitamin E. Thus, a review of vitamin E requires a comprehensive discussion of current multidisciplinary research, including information on free radical biology and lipid peroxidation of phospholipids in cellular and subcellular membranes containing polyunsaturated fatty acids (PUFA).

Vitamin E Nomenclature

The natural vitamin E of commerce (D-α-tocopherol; Fig. 1) is the most abundant and active isomer. It is prepared using tocopherols from vegetative oil distillates which contain a series of tocopherols and tocotrienols. There are eight different isometric forms of commercial synthetic vitamin E (α, β, γ, δ tocopherols and α, β, γ δ tocotrienols). The tocotrienols are similar to the corresponding tocopherols but with unsaturated side chains. These unsaturated isomers have not been well studied, and only α tocotrienol appears to have significant vitamin E activity. The biologic activity in international units per milligram is 1.49 for the natural source, and the synthetic preparations range between 1.00 and 1.21.

It is important for researchers to state which form of vitamin E preparation is being reported on, for clarity and uniformity in publications.

History of Vitamin E

Vitamin E was discovered in 1922, when Evans and Bishop demonstrated the existence of a fat-soluble factor which was necessary for normal reproduction in the rat (Evans 1962; Evans and Bishop 1922, 1923). Sure (1924) was the first to suggest the name vitamin E, which was soon adopted by Evans and Bishop.

In 1931, Olcott and Matill described the antioxidant function of vitamin E, which was subsequently recognised as its most important biologic function.

A variety of vitamin E deficiency syndromes produced experimentally in several animal species were described in 1931 (Pappenheimer and Goetsch 1931; Goetsch and Pappenheimer 1931).

In 1936, α-tocopherol was isolated by Evans et al. and the name tocopherol was suggested to Evans by George M. Calhoun, Professor of Greek at the University of California (Evans 1962; Parkes et al. 1969), where Evans worked. Tocopherol was coined from the Greek words "tocos," meaning offspring, and

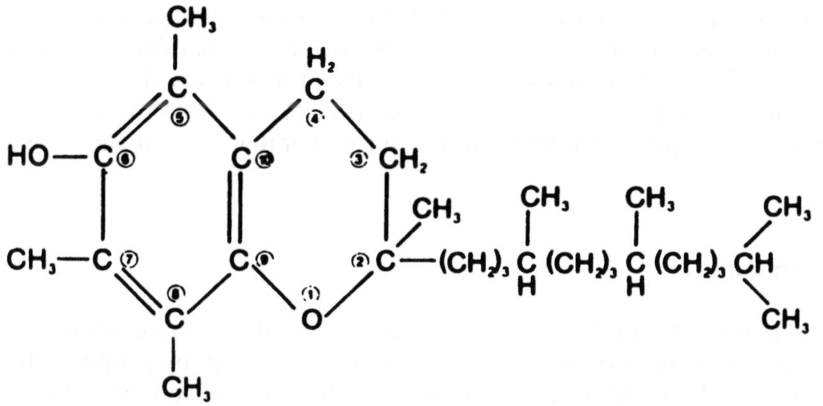

CHEMICAL NAME: 2, 5, 7, 8-Tetramethyl-2-(4', 8', 12'-trimethyl tridecyl)-6 chromanol

IUPAC NAME: 2R, 4'R, 8'R-Alpha Tocopherol

TRIVIAL NAME: RRR-Alpha Tocopherol

COMMON NAMES: d-Alpha Tocopherol
Natural Vitamin E

Fig. 1. Natural Vitamin E

"phero," meaning to bear. In 1938, Fernholz described the chemical structure of α-tocopherol and it was synthesized by Kerrer et al. in 1938. For the next nearly 50 years, there were reports of the efficacy of vitamin E in humans, from the newborn to the geriatric sector and cutting through all specialties. Many of the earlier reports dealt with nearly every conceivable medical disorder, ranging from aiding hair growth, skin disorders (used in lotions/creams), burns, intermittent claudication, prevention of aging, muscular dystrophy and sterility in men. Most of the earlier studies were uncontrolled for disorders with no proven vitamin E deficiency, and the explanations given were that the patients were acting as their own controls after trying various medications.

Current studies still reveal a large lacuna in our knowledge of this fascinating vitamin. The studies being reported still have controversial findings with no

Table 1. Clinical characteristics of patients

Characteristic	Study group (n = 12)	Control group (n = 12)
Sex	7 M/5 F	6 M/6 F
Age range	6–17 yr	6–17 yr
Mean age	10.8 yr	10.3 yr
Mean age of seizure onset	2.9 yr	2.6 yr
Mean duration of epilepsy	5.6 yr	5.5 yr

consistent reports. The reasons for this include inconsistent or unspecified types of vitamin E used and routes of administration, including solubilizer used in product dosage, timing of administration, previous plasma levels of vitamin E, sample size, and the stage of disease or problem studied. This list is by no means exhaustive, and one questions how reports can be compared with the above variables.

Vitamin E Therapy

Landmark experiments conducted in the 1940s and 1950s heralded numerous reports of prevention of various complications of prematurity by prophylactic administration of vitamin E. The diseases reportedly being prevented included bronchopulmonary dysplasia, retinopathy of prematurity and periventricular/intraventricular haemorrhage. However, the reports from various researchers to date have been controversial as far as efficacy in preventing any of these diseases is concerned.

In 1948, Gyorgy and Rose demonstrated that vitamin E-deficient rats developed haemolysis which could be corrected by tocopherol. In 1952, Gyorgy et al. showed that the erythrocytes of term infants were susceptible to haemolysis in the presence of hydrogen peroxide and that vitamin E supplements prevented this haemolysis. This susceptibility to haemolysis was the basis for the haemolytic anaemia that was described in the 1960s (Hassan et al. 1966; Oski and Barness 1967; Ritchie and Fish 1968; Williams et al. 1975; Bell and Filer 1981; Gerloczy 1949) in premature infants who were fed formulas that contained little vitamin E but abundant polyunsaturated fatty acids and iron.

Infants with severe fat malabsorption (abetalipoproteinaemia) and severe cholestasis are unable to absorb adequate vitamin E from the gastrointestinal tract and gradually develop a hemolytic anaemia and progressive spinocerebellar degeneration over a period of months or years (Sokol et al. 1985). The neurological symptoms can be both prevented and treated (partially) by the use of parenteral vitamin E (Guggenheim et al. 1982).

Vitamin E and Epilepsy

The brain mechanisms underlying epilepsy are not completely understood. In some experimental models, vitamin E has been found to reduce seizure activity (Jerrett et al. 1973; Wilmore et al. 1986). It prevented seizures from occurring by its administration in one model and significantly delayed the onset of electrographic seizures in the other model.

Clinical studies in children have found reduced plasma levels of vitamin E in epileptics on anticonvulsants but not in those not et receiving antiepileptic drugs or in age-related controls (Ogunmekan 1978a, b, 1979a, b, 1985; Nagura et al. 1985; Higashi et al. 1980).

Vitamin E inhibits the effects of oxidation in brain tissues (antioxidant) and is a free radical scavenger. Free radical scavengers have been implicated in the epileptic brain injury caused by prolonged seizures due to hyperoxia. Vitamin E is also a membrane stabilizer (Soderfeldt et al. 1983); protecting cell membranes from peroxidation damage.

An anticonvulsant action of vitamin E has been demonstrated in the intracerebral ferrous chloride model of epilepsy (Wilmore et al. 1986). An uncontrolled, non-blind clinical trial suggested that add-on vitamin E improved seizure control in some patients (Kovalenko et al. 1984).

In a recent study, we attempted to confirm the encouraging preliminary results by doing a prospective, modified crossover, randomized, double-blind, placebo-controlled, clinical trial of add-on vitamin E therapy (Ogunmekan and Hwang 1989). The hypothesis was that vitamin E deficiency exacerbated epileptic seizures in patients receiving antiepileptic drugs (AEDs), and treatment of the vitamin E deficiency should improve seizure control and/or neurologic function. The results from the first 24 patients who completed the double-blind, three-month period have been published (Ogunmekan and Hwang 1989) and are presented here.

Materials and Methods

Vitamin E in both capsule form and syrup preparation (each capsule or 5 ml syrup contained 294 mg of pure D-α-tocopheryl acetate in a solubilizer equivalent to 400 IU synthetic vitamin E, and the matching placebos were donated by Webber Canada Inc., Ontario, Canada).

Study Design and Treatment Schedule

Study subjects were drawn from > 500 epileptic children attending the Seizure Clinic of the Neurology Service at The Hospital for Sick Children in Toronto. Subjects had attended the clinic for at least 6 months, with a stable seizure pattern, stable AEDs, and at least four seizures per month.

Subjects were observed for an initial three-month baseline pre-trial period for familiarization with seizure documentation and for identifying and enrolling subjects in the study. The 24 randomly selected subjects ranged in age from 5 to 18 years. Subjects predominantly had generalized tonic-clonic seizures but could have other seizure types (e.g., complex partial, absence, myoclonic) that were clearly distinguished. Electroencephalographic (EEG) studies were performed using 16-channel recording with surface electrodes, according to the 10–20 system, with a recording time of at least 30 min using three montages: anteroposterior (AP) bipolar, coronal, and ipsilateral ear reference, together with intermittent photic stimulation, hyperventilation, and often spontaneous sleep. Biochemical parameters including liver enzymes, blood levels of AEDs, vitamin E, detailed seizure frequency record, and pill count were monitored throughout the study. Written informed consent was obtained (consent of parent was also required if patient was aged < 13 years old). The scientific protocol was approved by the Scientific Advisory Committee and the Human Subjects

Committee of the Research Institute at The Hospital for Sick Children, following guidelines of the Medical Research Council of Canada.

1. children aged <5 years who might be predisposed to febrile convulsions
2. patients with malabsorption or gastrointestinal disturbances that could interfere with vitamin E absorption
3. patients receiving iron therapy (including vitamins) or medications other than AEDs, because of the possibility of drug interactions, including haemolysis
4. patients anticipating or undergoing changes in AEDs, in whom seizure frequency might be variable during the study period
5. patients with adverse reactions to vitamin E or the solubilizer

Study Procedure

Vitamin E (294 mg) or matched placebo was administered as a single daily dose for 3 months, in addition to the patient's usual AEDs. A study code was maintained by The Research Pharmacy Department and investigators, patients, and parents were kept "blind" to the code. At 3 months, the code was broken. Responders receiving vitamin E had the option to continue. Those receiving placebo were crossed over to active vitamin E for 3 months as an open study. The study code drug could be broken in an emergency if a participant developed status epilepticus or any otherwise unexplained adverse reaction.

Parents kept daily records of the frequency of the child's seizures between hospital visits and returned unused vitamin E or placebo for pill counts or measurement of liquid preparation at each hospital visit. Steady-state plasma levels of AEDs and vitamin E and haematological parameters including haemogram, platelets, SGOT, alkaline phosphatase, and amylase were determined at each visit. Vitamin E or matching placebo was usually taken between 16 and 20 h prior to venipuncture and the last dose(s) of AED was taken 6–8 h prior to venipuncture. The hospital visits study procedure is shown in Fig. 2.

Minor changes, such as alteration in times of AEDs or individual doses without altering the *total* daily dosage, were permissible to ensure compliance with ethical concerns. Major changes such as withdrawal or introduction of a new AED were not permitted during the study.

At the end of the study the effects on seizure frequency between the study and control groups in the double-blind phase and between the placebo and active vitamin E in the open-label phase were compared. Data were also examined to determine if effectiveness was maintained or if further improvement occurred in patients continuing with vitamin E for 3 months after the initial beneficial effect.

Vitamin E Analytical Procedure

Serum α-tocopherol (vitamin E) levels were determined by the micromethod of Hashim and Schuttringer (1966). β-carotene interference at the wavelength at

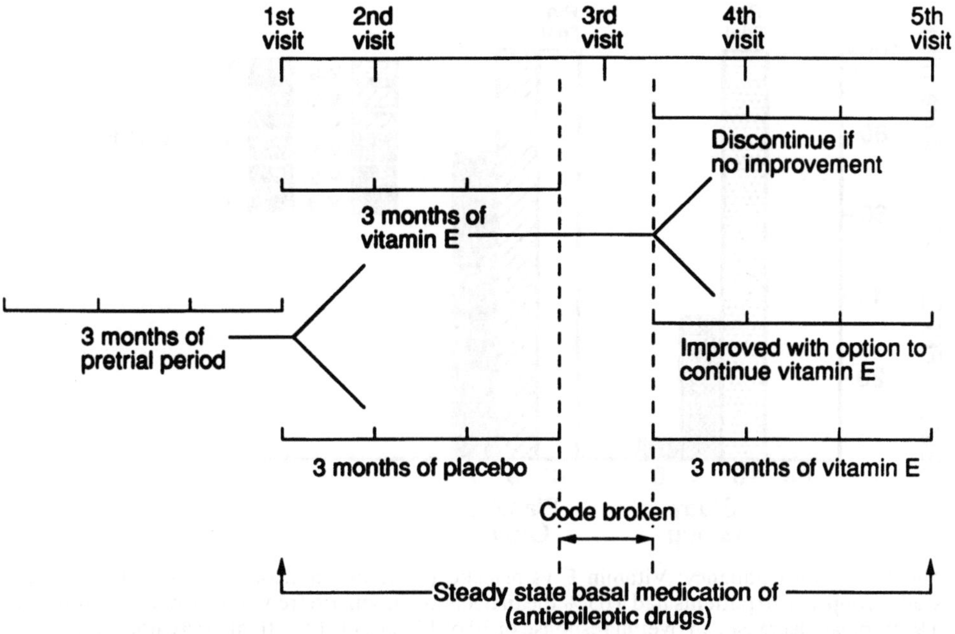

Fig. 2. Study procedure. Modified cross-over design comparing D-α-tocopherol (vitamin e) to placebo in a randomized double-blind study (first phase), followed by an open-label study in which patients receiving placebo originally served as their own controls

which tocopherol was measured was calculated and substracted to obtain the α-tocopherol concentration. Analyses were performed at the Haematology Laboratory of The Hospital for Sick Children and were not made available to the "blind" investigators throughout the study period.

EEG

EEG studies were interpreted blind by one of the investigators (P.H.) and graded on a 5-point scale for background activity and epileptiform activity (1, normal; 2, mild; 3, moderate; 4, marked; 5, severe).

Results

Seizure Response

Of 12 subjects receiving active drug, ten were considered responders (i.e., > 60 % reduction in seizure frequency); six had 90 %–100 % reduction of seizure frequency, and four had 60 %–90 % reduction. Two were considered failures (< 60 % reduction in seizure frequency). The two failures were clinically

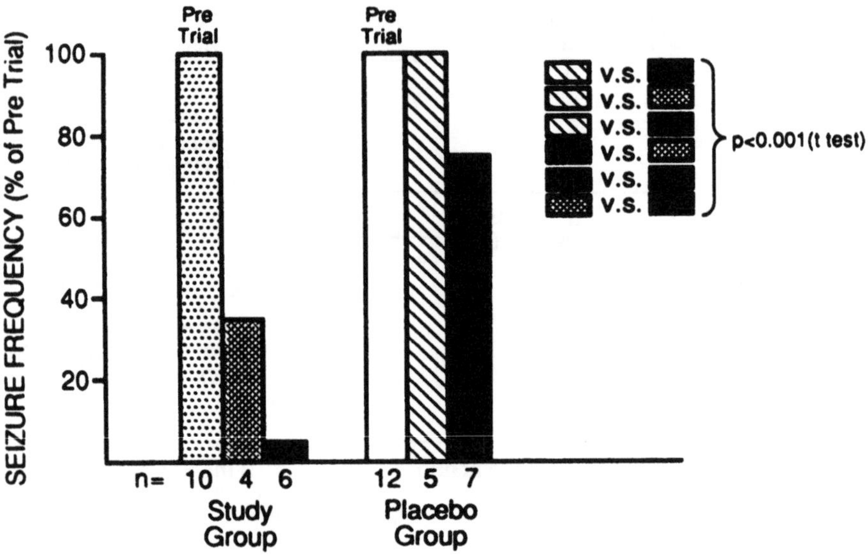

Fig. 3. Seizure frequency: Vitamin E vs placebo (excluding noncompliant patients). In the study group 4 of 10 patients had a 65 % mean decrease in seizure frequency, whereas 6 of 10 had a 95 % mean decrease, an overall response of 10 of 12 patients (2 patients were noncompliant). Significant difference from placebo group (7 of 12 with 25 % decrease in seizure frequency and 5 of 2 with no change), $p < 0.05$ (t-test)

identified as noncompliant subjects during the trial period, a status that was also confirmed by vitamin E assays. Of 12 subjects receiving placebo, 42 % (5 of 12) had little change in seizure frequency; 58 % (7 of 12) had 25 % reduction in seizure frequency (Fig. 3). They are all considered "failures" by the criteria of < 60 % seizure reduction. The proportion of responders in the active treatment group (10 of 12 or 83 %) was significantly higher than that in the placebo group (0 of 12 or 0 %; $p < 0.05$).

Which Variables Influence the Clinical Effect of Vitamin E?

Many clinical variables were tested; none was predictive of response. In particular, the clinical effect of vitamin E did not appear to be influenced by age, sex, duration of epilepsy, number of seizure types, type of epilepsy, or neurological deficit. The EEG findings, however, appeared to have improved in > 50 % (4 of 7 patients) of the responders who had EEG performed before and after trial of add-on vitamin E. Three of the responders had normalized background activity with abolition of epileptiform disturbances (grade 1). One responder had improvement of background activity only, one responder deteriorated from mild (grade 2) to moderate epileptiform (grade 3) disturbances, and two showed no change in either background activity or epileptiform disturbances.

Comedications

Of ten subjects showing response while receiving vitamin E, four were receiving one AED, five were receiving two, and one was receiving three. In the placebo group, five were receiving one AED, one was receiving two, and one was receiving three. There was no appreciable difference in plasma levels of carbamazepine, valproate, or phenytoin in vitamin E responders with respect to the entire group of patients as compared with the pretrial phase.

Seizure Types

Of ten vitamin E respondent patients two had generalized tonic-clonic seizures only; three had generalized tonic-clonic, complex partial, and simple partial seizures; one had complex partial seizures only; one had complex partial seizures and generalized tonic-clonic seizures; one had complex partial seizures with myoclonic and atonic seizures; and one had generalized tonic-clonic seizures with myoclonic and absence seizures. One had generalized tonic-clonic and simple partial seizures. In the control group, comparable numbers of seizure types were also represented.

Clinical Laboratory Tests

No clinically significant treatment-related alterations of haematologic or biochemical laboratory test results were noted during add-on vitamin E therapy. One subject receiving valproate during the trial had a moderate elevation of serum aspartate aminotransferase (glutamic oxaloacetic-transaminase) and serum amylase. These tests were normal when repeated in 2 weeks.

Another patient developed status epilepticus during the trial period after developing chicken pox. This patient was removed from the final analysis of results.

Discussion

Previous studies (Kovalenko et al. 1984) showed therapeutic efficacy of add-on therapy of 600 mg daily of vitamin E in a therapy-resistant epileptic population; the study was, however, not blind. Our double-blind, placebo controlled, add-on trial showed a significant improvement in seizure control in therapy-resistant epileptic children when the active group was compared to the placebo group.

Clinical efficacy demonstrated in the present study was not due to an interaction with the co-medications, as the blood levels of AEDs were not significantly altered. Plasma vitamin E level increased with a daily dose of 294 mg ranging from 5.3 to 36.9 μM, with a mean rise of 24.5, except in the noncompliant subjects and during the placebo phase. The study also suggests that generalized tonic-clonic seizures will probably respond to vitamin E, but this observation is confounded by the predominance of tonic-clonic seizures. The marked variation of rise in vitamin E levels in responders may have been due to individual variability in body size, volume of distribution, protein binding or other pharmacokinetic characteristics. Whether vitamin E in a higher dose or in a less therapy-resistant population would be capable of providing complete seizure control or whether it may prove to be a major AED or merely an adjuvant cannot be predicted. In recent literature review, EEG was considered to be of doubtful value as an outcome variable in clinical AED trials (Van Wieringen et al. 1987).

Our study is being expanded with new patient entry. The response ofthe double-blind phase and the open-label phase for the subjects originally receiving placebo for the first 3 months showed similar findings (reduction in seizure frequency between 70 % and 100 % in all patients), and the response of the six subjects continuing with vitamin E after beneficial effects in the first phase showed continued improvement.

Conclusions

The findings in this study would justify a larger, longer Phase III trial to confirm that vitamin E supplements may have a place as an add-on therapy in seizure control. This therapeutic observation accords with previous reports from uncontrolled observations and from experimental studies. This may be very important to the 10 %–20 % of the epileptic population whose seizures are inadequately controlled with the best anticonvulsant drugs presently available.

Acknowledgement. We are grateful to the parents and children who participated in this study; Lyle Merrell, President, Webber, Canada for financial support and supplier of D-α-tocopherol acetate preparation and placebo; R.E. Hudson, Vice President, Eastman Chemical Products, Tennessee for financial and material support; The Hospital for Sick Children Foundation for a Visiting Scholar award to Dr. Ogunmekan; Dr.S ue Phancharoen, Chulalongkorn University, Bangkok, Thailand; Dr. W.J. Logan, Head of Neurology, Dr. Margaret Wood, biostatistician, and Dr. Annete Poon, Division of Haematology, The Hospital for Sick Children.

References

Bell EF, Filer IJ Jr (1981) The role of vitamin E in the nutrition of premature infants. Am J Clin Nutr 34: 414–422

Evans HM (1962) The pioneer history of vitamin E. Vitam Horm 20: 379–387

Evans HM, Bishop KS (1922) On the existence of a hitherto unrecognized dietary factor essential for reproduction. Science 56: 650–651

Evans HM, Bishop KS (1923) The production of sterility with nutritional regimens adequate for growth and its cure with other foodstuffs. J Metab Res 3: 233–316

Evans HM, Emmerson DM, Emmerson GA (1936) The isolation from wheat germ oil of an alcohol α-tocopherol having the properties of vitamin E. J Biol Chem 113: 329–332
Fernholz E (1938) On the constitution of α-tocopherol. J Am Chem Soc 60: 700–705
Gerloczy F (1949) Clinical and pathological role of d, L, α-tocopherol in premature infants (studies on the treatment of scleroderma). Ann Paediatr (Basel) 173: 271–286
Goetsch M, Pappenheimer AM (1931) Nutritional muscular dystrophy in the guinea pig and rabbit. J Exp Med 54: 145–165
Guggenheim MA, Ringel SP, Silverman A et al. (1982) Progressive neuromuscular disease in children with chronic cholestasis and vitamin E deficiency: diagnosis and treatment with α-tocopherol. J Pediatr 100: 51–58
Gyorgy P, Rose CS (1948) Effect of dietary factors on early mortality and haemoglobinuria in rats following administration of alloxan. Science 108: 716–718
Gyorgy P, Cogan G, Rose CS (1952) Availability of vitamin E in the newborn infant. Proc Soc Exp Biol Med 81: 536–538
Hashim SA, Schuttringer GR (1983) Rapid determination of Nocopherol in macro and micro quantities of plasma. Am J clin Nutr 19: 137
Hassan H, Hashim SA, Van Itallie TB, Sebrell WH (1966) Syndrome in premature infants associated with low plasma vitamin E levels and high polyunsaturated fatty acid diet. Am J Clin Nutr 29: 147–147
Higashi T et al. (1980) Serum vitamin E concentrations in patients with severe multiple handicaps treated with anticonvulsants. Paed. Pharmacol 1: 129–134
Jerrett SA·et al. (1973) Seizures H_2O_2 and lipid peroxidase in brain during exposure to oxygen under high pressure. Aerosp Med 44: 40–44
Kerrer P, Fritzsche H, Riesgier GH, Solomon H (1938) α-Tocopherol. Helv Chim Acta 21: 520–525
Kovalenko VM et al. (1984) Alphatocophenol in the complex treatment of several forms of epilepsy. Neuropathol Psikhitu 6: 892–897
Nagura et al. (1985) Decreased serum vitamin E levels and increased serum peroxide in epileptic children. 16th epilepsy international congress
Ogunmekan AO (1978a) Is there a role for vitamin E therapy in epilepsy in man? MD thesis, University of Newcastle
Ogunmekan AO (1978b) Vitamin E deficiency in animals and man. Can J Neurol Sci 6: 43–45
Ogunmekan AO (1979a) Relationship between age and vitamin E levels in epileptic and normal children. Am J Clin Nutr 32: 2269–2271
Ogunmekan AO (1979b) Predicting serum vitamin E concentrations from the age of normal and anticonvulsant drug treated epileptic children using regression equations. Epilepsia 20: 295–301
Ogunmekan AO (1985) Plasma vitamin E levels in apparently normal children and epileptic children with and without anticonvulsant drug therapy. Trop Geog Med 37: 275–277
Ogunmekan AO, Hwang PA (1989) A randomized, double-blind, placebo-controlled, clinical trial of D-α-tocopheryl acetate (vitamin E), as add-on therapy, for epilepsy in children. Epilepsia 30/1: 84–89
Olcott HS, Matill HA (1931) The unsaponifiable lipids of lettuce. Fractionation. J Biol Chem 93: 59–64
Oski FA, Barness LA (1967) Vitamin E deficiency: a previously unrecognized cause of haemolytic anaemia in premature infants. J Pediatr 70: 211–220
Pappenheimer AM, Goetsch MA (1931) A cerebellar disorder in chicks, apparently of nutritional origin. J Exp Med 53: 11–26
Parkes AS, Simpson M, Lyons WR (1969) Dr. H.M. Evans an interview. J Reprod Fertil 79: 3–19
Ritchie JH, Fish MB (1968) McMasters: V Grossman, M Oedema and haemolytic anaemia in premature infants: a vitamin E deficiency syndrome. N Engl J Med 279: 1185–1190
Soderfeldt B, Blennow G, Kalimo H et al. (1983) Structural changes accompanying bicuculline induced seizures in rats following manipulations of tissue oxygenation or alphatocopherol levels. Acta Neuropathol (Berl) 60: 80–91

Sokol RJ, Guggenheim MA, Heub JE et al. (1985) Frequency and clinical progression of vitamin E deficiency neurological disorder in children with prolonged neonatal cholestasis. Am J Dis Child 139: 1211–1215
Sure B (1924) Dietary requirements for reproduction. The existence of a specific vitamin for reproduction. J Biol Chem 58: 694–709
Van Wieringen A, Binnie CD, Overweg J et al. (1987) Electroencephalographic findings in antiepileptic drug trials: a review and report of 6 studies. Epilepsy Res. 1: 3–15
Williams ML, Short RJ, O'Neal PL, Oski FA (1975) Role of dietary iron and fat in vitamin E deficiency and anaemia of infancy. N Engl J Med 292: 887–890
Wilmore J et al. (1986) The role of iron induced hippocampal peroxidation in acute epileptogenesis. Brain Res 382/2: 422–426

Subject Index

MIX
Papier aus verantwortungsvollen Quellen
Paper from responsible sources
FSC® C105338

If you have any concerns about our products,
you can contact us on
ProductSafety@springernature.com

In case Publisher is established outside the EU,
the EU authorized representative is:
Springer Nature Customer Service Center GmbH
Europaplatz 3, 69115 Heidelberg, Germany

Printed by Libri Plureos GmbH
in Hamburg, Germany